Perfume in the Bible

Perfume in the Bible

Charles Sell
Email: charles.s.sell@gmail.com

For my wife Hilary

to whom I am especially grateful for her support, her good humour and patience when I shut myself away in the study for hours on end working on the book.

Print ISBN: 978-1-78801-730-5
EPUB ISBN: 978-1-78801-838-8

A catalogue record for this book is available from the British Library

The Royal Society of Chemistry is a charity, registered in England and Wales, Number 207890, and a company incorporated in England by Royal Charter (Registered No. RC000524), registered office: Burlington House, Piccadilly, London W1J 0BA, UK, Telephone: +44 (0) 207 4378 6556.

Visit our website at www.rsc.org/books

Printed in the United Kingdom by CPI Group (UK) Ltd, Croydon, CR0 4YY, UK

Foreword

The daily prayer of the Church of England has for centuries set psalms for morning and evening prayer for each day of the month so that those who are used to the rhythm of worship, either alone or in community, find themselves coming to know verses of the psalms and the character of the psalms very well. The psalms are, of course, a gift from the Hebrew Bible and are poetry that we have in common with other faiths and denominations. I find that, on particular mornings, my memory and physical appreciation of the psalms comes before the understanding of what I am saying, so that in, for example, Psalm 45, well before we recite verse 9, a pleasant anticipation of fragrance is all around me. For verse 9 says of the king "All thy garments smell of myrrh, aloes, and cassia: out of the ivory palaces, whereby they have made thee glad". At that moment, the scent and physical sense of what is being said become more powerful than the mental understanding, in the same way as the frankincense and myrrh of the story of the coming of the Magi give that same sensation in St Matthew's Gospel. It is the essence of perfume to awaken the senses and, as Charles Sell has said in

Perfume in the Bible
By Charles Sell
© Charles Sell 2019
Published by the Royal Society of Chemistry, www.rsc.org

this book on perfume in the Bible, the scriptures of both the Old Testament and the New Testament are full of references to scents and the power of smell in human life to awaken the mind and heart in an astonishing way.

I found on reading this book that I was on an amazing and exciting journey into areas that I knew a little about and, through constant reading of the scriptures, was alive to, but the depth and interconnection of the fragments of my knowledge began to be ordered by the various facts and ideas given in the chapters one by one. We are taken on a journey through different kinds of disciplines through the history of the ancient world, through scientific discovery, through studies in plants and trees and into theological and ecumenical areas, all of which add to the pleasure of Charles Sell's fascinating story. We see how ancient languages and ancient cultures feed into each other and how perfumes from the Far East and China are imported into the eastern Mediterranean region many years before the stories we are used to in our Bibles are recounted.

To those of us who know the scriptures well or the classical writers in any way at all, we find ourselves in familiar places with familiar people but re-interpreted according to the questions that Charles Sell is asking as to how perfumes were created and what their intention was. Finally, we find ourselves amongst the Gospel writers and, particularly, with the story common to all four Gospels of the anointing of Jesus by a woman who breaks open an alabaster jar of precious ointment, the perfume of which fills the house. The book gives many answers and provokes many questions but it never fails in fascination and is beautifully illustrated with pictures showing the way in which incense and perfumes were used throughout the centuries, both to honour God and give delight and well-being to human life. It gave me great pleasure to read it and gives me great pleasure to be able to write a foreword introducing it.

Robert Willis

Dean of Canterbury

Preface

My research into the subject of Biblical perfumes began when I was invited by a friend, Dr Ted Harrison, to give a lecture on a subject linking science to the Bible. I am grateful to Ted for setting this project in motion. The subject proved very popular and my talk has grown and evolved over time. Audience members invariably ask if I could recommend a book on the subject and this led me to the realization that there were none. So, I decided to take on the task of writing one. Whilst training as a volunteer guide at Canterbury Cathedral, I found that its beautiful medieval windows included many that illustrated people and events relating to perfume, and thus the idea of using them to illustrate this present book was born.

My aim in writing the book is to fill a gap and to show how interrelated are the many diverse subjects involved, including chemistry, botany, sensory science, medicine, Biblical studies and the arts. The book is intended for a wide readership and to be accessible to everyone, whether familiar with any of the above disciplines or none. For this reason, terms that are essentially jargon to a specific discipline are printed in **boldface** the first time they are used and an explanation will be found in the glossary.

When citing dates in history, I use the abbreviations BC (Before Christ) and AD (Anno Domini) rather than the alternatives BCE (Before Common Era) and CE (Common Era) which

Perfume in the Bible
By Charles Sell
© Charles Sell 2019
Published by the Royal Society of Chemistry, www.rsc.org

are commonly used in books on Biblical studies. The use of BCE and CE illustrate the unity of the Old and New Testaments of the Christian Bible and the fact that Christianity is founded on Judaism. My reason for using BC and AD is that these terms are more widely understood by those outside the two religions. The book is written from a Christian perspective but I most certainly do not intend any offence to those of other faiths or none. Scientific information is true to the best of my knowledge. I have taken Biblical statements at face value and the reader must decide how (s)he should understand them.

Phonetic representations of Hebrew words are as given on the Biblehub website. For Greek words, I have used simple transliteration of Greek into English. In order to distinguish between the two e and o sounds as per the generally accepted pronunciation for Greek of the first century, I have used e to denote pronunciation as in the English word peg and ę when it should be pronounced as in the English word deep. Similarly, o represents the sound in the English word pot and ǫ the sound in the English word pole. The Greek ch is hard, as in the Scottish pronunciation of loch.

The original Biblical texts were written in blocks of text as can be found, for example, in a modern book or newspaper. It was Archbishop Stephen Langton of Canterbury who devised a scheme of dividing Biblical books into chapters (usually representing paragraphs in the originals) and verses (usually sentences in the originals) to make it easier to find specific parts of the text. So, when citing Biblical references, I use his system and present citations in the format of Book, chapter.verse(s).

References to articles in scientific journals are numbered using superscript numerals and the journal references will be found at the end of each chapter. References to books are by book title in *italicised script* and details of the books will be found, among others, in the bibliography.

In writing the book, I have had help from a number of people and wish to express my gratitude to them. My brother-in-law, Kevin Stock, gave me great help both with word searches and with Hebrew, and I am also grateful to Professor Simon Majaro, MBE, for valuable insights into Hebrew and to Professor Michael Abraham who introduced me to him. I would also like to thank several former colleagues in perfumery for their input from the point of view of that art: Chris Sheldrake, Roger Duprey, David Hooper and Brian Jaggers have all been helpful. My thanks also go to Professor John Butler and Elizabeth and James Mann for the encouragement they gave me that helped me persevere with the project. I am very grateful to the Dean and Chapter of Canterbury for permission to use the window images, to the Dean, the Very Reverend Robert Willis, for his support and encouragement, to Chris Needham of Cathedral Enterprises, Caroline Plaisted of the Friends of Canterbury Cathedral and Chris Pascall of the Visits Office for their support and particularly to Léonie Seliger the Cathedral Glazier for her expert advice and generous help with the window images. Some other images were given to me by my former colleagues in Givaudan, Dr Robin Clery and Dr Roman Kaiser, by Givaudan and by Dr Rui Fang of Kew Gardens, so my thanks go to all of them. Specific acknowledgements for their photographs will be found in the figure captions. My thanks also go to Philip Oostenbrink (Head Gardener at Canterbury Cathedral) and my friend Helen Tregenza Crofts for helping me find growing specimens of *Liquidambar orientalis* and *Styrax japonica*. Naturally, my thanks also go to Janet Freshwater and Katie Morrey at the Royal Society of Chemistry for all their help in bringing the book to publication.

I am also very grateful to Givaudan for their generous donation to cover the cost of producing the Canterbury Cathedral images and to my former colleagues Jeremy Compton, Marie-Laure Andre and Victoria Joseph for their support and help with other photographs.

Charles Sell

Glossary

Terms included in the glossary are indicated in bold upon their first appearance in the book.

Absolute – A perfume material which was first extracted from plant material using fat, then the odorous components were extracted from the fat using alcohol and, finally, the alcohol was removed by evaporation or distillation.

Aldehyde – This is a word used to describe molecules in which one carbon atom is bonded to one other carbon atom and one hydrogen atom and doubly bonded to an oxygen atom. In perfumery, the term usually implies a fatty aldehyde; that is, a long chain molecule containing no atoms other than carbon and hydrogen in the molecule, apart from one oxygen atom doubly bonded to a carbon atom at the end of the chain.

Aliphatic hydrocarbon – A chemical, the molecules of which contain only carbon and hydrogen atoms and no rings.

Analgesic – A substance that counteracts pain.

Anneal – The process of heating a glass object made from fusing different pieces of glass in order to ensure that the components become one single unit.

Anti-type – See **typology**.

Carotenoid – A chemical belonging to the terpenoid family (see below) that contains 40 carbon atoms, usually either in a chain or with one ring at each end. They are natural pigments, a

Perfume in the Bible
By Charles Sell
© Charles Sell 2019
Published by the Royal Society of Chemistry, www.rsc.org

typical example being the carotenes that give carrots their characteristic orange colour.

Concrete – A fatty or waxy substance produced by extraction of a plant using a solvent. If the solvent is volatile, it would be removed by distillation to give the concrete. If the solvent is a fat, the mixture of fat and odorous material is also referred to as a concrete.

Core formed – A glass bottle made by coating an earthenware core with molten glass then, once the glass has solidified, removing the earthenware core.

Enfleurage – A process involving extraction of odorous plant components into a layer of purified fat.

Enzyme – A protein that is responsible for carrying out chemical reactions in the cells of plants, animals or micro-organisms.

Essential oil – A volatile oil produced by distillation of either plant material or an extract from a plant. Originally called a quintessential oil because it was believed that it was the quintessence (fifth element), that is, the spirit, of the plant.

Evangelist – See **gospel**.

Gospel – The modern English word gospel is derived from the Anglo-Saxon "good spell", meaning good news, and is a direct translation of the Greek word ευαγγελιον (euangelion). Gospel can mean either one of the first four books of the New Testament or the news they contain. The word evangelist is derived from the original Greek word and means the writer of one of the four gospels. It is sometimes also extended to anyone who spreads the good news.

Kohl – Kohl was made in ancient Egypt by subliming lead sulfide (PbS), in other words, heating it to vaporise it and then cooling the vapour to produce a fine black powder. It was used as eye shadow by the Egyptians.

Lipid – A fatty substance, either solid or liquid; usually one comprising long chain fatty acids or derivatives of these. The fatty acids are a group of natural chemicals. Their molecules contain a long chain of carbon atoms with a carboxylic group at the end of the chain. A carboxylic acid group comprises one carbon atom bonded to one other and also to two oxygen atoms, one by a double bond and the other by a single bond to an oxygen atom that is also bonded to a hydrogen atom.

Monomer – A small molecule that can link to other small molecules to form a long chain, called a polymer. For example, styrene is a small molecule that can bind to other styrene molecules to form the polymer, polystyrene.

Monoterpene – See **terpene**.

Monoterpenoid – See **terpenoid**.

Mould blown – A glass bottle made by blowing a bubble of molten glass into a mould. When the glass has hardened, the mould is removed to release the bottle.

Narthex – A room or space on the Western side of a Christian church that was accessible for people who did not have the right to enter the church. For example, in the Middle Ages, a monastic abbey church was expressly for the use of the religious community but others could enter as far as the narthex.

Natron – This term refers to natural deposits of sodium carbonate decahydrate ($Na_2CO_3 \cdot 10H_2O$, washing soda), usually containing some sodium bicarbonate ($NaHCO_3$, baking soda) and other impurities.

Nature identical – A chemical substance made using organic chemistry in a laboratory or factory and which is identical to the same substance made in plants, animals or micro-organisms.

Neuroprotective – Something that protects nerves from damage.

Oculus – A circular window that has no stonework inside the circle.

Oleoresin – A plant exudate comprising a volatile oil and an involatile resin.

Percept – A mental image constructed by the brain using input from sensory nerves (those conveying signals of light, sound, odour molecules, taste molecules, temperature, pressure *etc.*, from outside the body) together with input from higher brain regions such as those responsible for experience, expectancy and context.

Polymerise – The chemical process in which small molecules join together to form a long chain. See **monomer**, above.

Polysaccharide – A substance, such as starch, whose molecules are made up from sugar molecules.

Pyrolysis – The chemical processes by which molecules are broken down to smaller units by heat.

Quintessential oil – See **essential oil**.

Resinoid – A substance produced by extraction of plant material using a solvent, usually a volatile solvent, and containing a volatile oil and/or an involatile resin and/or an involatile gum.

Sesquiterpenoid – See **terpenoid**.

Shikimate – Chemicals produced by living organisms in a process using the chemical known as shikimic acid as an intermediate in the overall process are known as shikimates.

Shikimic acid – See **shikimate**.

Skin sensitisation – The body's immune system recognises and eliminates foreign substances. Sometimes an over-reaction occurs. Skin sensitisation is an example of this. Someone's skin might be exposed to a chemical substance (natural or synthetic) and their immune system learns to recognise this but over-reacts on subsequent exposure to produce a rash.

Synoptic gospels – The word synoptic is derived from Greek and means seeing together. It is a term applied to the first three gospels; Matthew, Mark and Luke. These gospels treat the material in a similar style in that they report the events of Jesus' life in a factual, almost journalistic, way. The fourth gospel, John, is more reflective and rather than recording many different events, picks out a few of them and elaborates on their theological significance.

Terpene – A member of the **terpenoid** (see below) family of chemicals. Strictly speaking, a monoterpene is one whose molecules contain 10 carbon atoms, the only other atoms present in the molecule being hydrogen. The term is sometimes misused to describe any monoterpenoid.

Terpenoid – This term refers to large group of chemicals produced by living organisms and for which a five-carbon molecule called isoprene is the key building block. Terpenoids containing two isoprene units are known as **monoterpenoids** and contain 10 carbon atoms arranged in a recognisable pattern in the molecule. Terpenoids containing three isoprene units, and hence 15 carbon atoms, are known as **sesquiterpenoids.** The boiling points and solubility of monoterpenoids are in a range that makes them ideal for transport through the air from the plant or perfume source to the nose and hence to stimulate the sensory proteins in the nose. Monoterpenoids are key contributors to the head and heart notes of many flowers and floral perfumes. Sesquiterpenoids have higher boiling points and so are less capable of reaching the nose. Those that do stimulate the olfactory system tend to do so quite strongly. The scent of many of the Biblical perfume ingredients results from stimulation of the receptors in the nose by sesquiterpenoids. There are many families of terpenoids with larger numbers of isoprene units (for example, steroids with 30 carbons and carotenoids with 40) but these have much higher boiling points and hence are not volatile enough to reach the nose. However, decomposition of these can produce smaller fragments that do reach the nose. For example, safranal is the molecule responsible for eliciting the odour of saffron and it is a degradation product of the carotenoid pigment responsible for the red colour of the saffron stamens.

Tetramorph – A form of symbolism that has origins in early Sumerian culture. In medieval Christianity, it came to represent the four evangelists. All four figures are winged: a man represents Matthew, a lion represents Mark, a bull represents Luke and an eagle represents John.

Tinctures – A solution in alcohol (ethanol) of an extracted plant material. The alcohol could also be the solvent used for extraction from the natural source. The whale excretion ambergris can also be used in perfumery in the form of a tincture.

Tympanum – The semi-circular construction between a lintel and the arch above it, usually of stonework, and often found in the doorways of medieval churches, abbeys and cathedrals. In the latter, they are usually decorated with elaborate carvings.

Type – See **typology**.

Typology – Typology is a system of relating events or objects that took place or originated in different historical times. The earlier event/object is called the **type** and the later one the **anti-type**. For example, in Gilbert and Sullivan's operetta "Iolanthe", the fairy queen sings about the "amorous dove" being a "type of Ovidius Naso". Thus, she indicates that the love poetry of the Roman poet Ovid follows the example of doves, a symbol of love that were there long before Ovid's birth. In Biblical terms, the types are Old Testament events that foretell those of the New Testament, the anti-types.

Umbelliferate – Plants whose flowers spring out from a central source giving an effect similar to the ribs of an umbrella.

Contents

Perfume in the Bible
By Charles Sell
© Charles Sell 2019
Published by the Royal Society of Chemistry, www.rsc.org

CHAPTER 1

Introduction

1.1 THE LANGUAGES OF THE BIBLE

The Christian Bible is a collection of 66 books and letters written over a considerable period of time. The Old Testament is the Jewish canon of scripture, containing the 5 books of the law (the Torah), 12 history books, 5 poetry books and 17 books of prophecy. These distinctions are somewhat arbitrary as, for example, there are historical accounts in some of the law books and the prophets. The New Testament is the Christian canon of scripture and contains four **gospels**, one history book, 21 letters and one book of prophecy.

The Old Testament was written mostly in Hebrew, with some parts of the book of Daniel written in Aramaic, which was a common language of the Babylonian court. The Kingdom of Israel divided into two parts in about 928 BC when Rehoboam succeeded his father Solomon as king and Jeroboam, son of Nabat, broke away forming a Northern kingdom initially known as Israel but later called after its capital city, Samaria. The Southern kingdom became known as Judah after the land allotted to the tribe of Judah on division of the Promised Land. In about 720 BC, the Assyrians invaded Samaria and it became an Assyrian province until the Babylonian invasion. The Babylonians under Nebuchadnezzar conquered both Samaria

Perfume in the Bible
By Charles Sell
© Charles Sell 2019
Published by the Royal Society of Chemistry, www.rsc.org

and Judah, capturing and destroying Jerusalem on 16th March 597 BC. Nebuchadnezzar deported many Jews to Babylon and both Samaria and Judah became Babylonian provinces. The Persian, Cyrus the Great, conquered Babylon in October 539 BC. He employed a more devolved system of government and allowed the Jewish exiles to return to their land, both Samaria and Judah becoming Persian satrapies, and encouraged the rebuilding of Jerusalem. The Persian Empire was, in turn, over-thrown by the Greeks under Alexander the Great in 331 BC. Alexander built a Greek Empire that stretched from Spain to India, and Greek became the common language of that Empire. On Alexander's death on 10th June 323 BC, there was a struggle for succession and his empire fragmented. Egypt came under control of the Ptolemies and Syria, including the Holy Land, under the Seleucids. By this time, Jews were living in many parts of what had been Alexander's empire and some were beginning to lose the ability to read Hebrew. It is said that the first trans-lation of the Jewish Bible into Greek was requested by Ptolemy II in Egypt. Modification and refinements of the translations were made, possibly to bring them closer to the Hebrew. The Septuagint is a translation that dates from the 2nd century BC and manuscripts of the Septuagint were found in the cave at Qumran (The Dead Sea Scrolls). This version of the Jewish Bible was the one in common use by Hellenised communities and is the version used in New Testament citations of the Old Testament. However, the Septuagint contains more material than there is in the Hebrew Bible. The additional 15 books and parts of books are nowadays known as the Apocrypha. The Apocrypha is not recognised as canonical scripture by either Judaism or Protestant Christianity and, although some refer-ences to perfume from it are used here, it was not a major part of the research for this present book. In the 2nd century BC, Judas Maccabeus and his brother led a Jewish revolt against Seleucid rule and invited the Romans to set up a garrison in Jerusalem to protect them from the Seleucids. By New Testa-ment times, the Roman Empire had replaced the Greek as the major power in Europe and the Middle East. However, Greek continued to be the common language of the Eastern part of the Roman Empire, and all educated Romans could read Greek. Therefore, although the native tongue of all but one of the writers of the New Testament would have been Aramaic, they wrote in Greek. Luke, who wrote the third gospel and the Acts of the Apostles, was the only contributor whose native language was Greek.

The implications for this present work are significant. Searches in Hebrew (Old Testament), Greek (Septuagint and New Testament) and English (various translations) identified words associated with perfume, perfumers, scents and the sense of smell but comparisons between translations are not always consistent. For example, the Hebrew word רֹקֵחַ (rō·w·qê·aḥ) is found in Exodus, 30.35; 1 Samuel, 8.13; and Nehemiah, 3.8. In modern Hebrew, it means pharmacist. In Exodus and 1 Samuel, the Greek of the Septuagint gives μυρεψικος (murepsikos), which means perfumer, and in Nehemiah, the Septuagint simply gives a phonetic representation of the Hebrew word. In some English translations, we find apothecary and, in others, perfumer. A discussion of this and similar problems of translations will be found in appropriate places in this book.

1.2 PERFUME AND THE BIBLE

Perfumery is one of the oldest industries and is mentioned in the writings of all ancient civilisations; therefore, it is not surprising that we also find it mentioned in the Bible. In fact, there are references to perfume in 24 of the 39 books of the Old Testament and in 9 of the 27 books of the New Testament. Thus, 61% of the books of the Old Testament mention perfume, as do 33% of those of the New Testament. In total, 33 of the 66 books of the Christian Bible mention perfume. In other words, perfume is included in 50% of the books of the Bible.

The first reference to perfume is right at the beginning of the Bible. In Genesis, 2.12, we are told that the land of חֲוִילָה (ḥă·wî·lāh) (Havilah) is a source of aromatic resins. Havilah is probably the Horn of Africa, an area which remains, to this day, a major source of resins such as frankincense. The same sentence in Genesis also refers to gold and so this could be an example of **typology** since "aromatic resin" could include frankincense and myrrh, and hence the text could foreshadow all three of the gifts brought by the magi to the infant Jesus (Matthew, 2.11).

The last reference to perfume comes right at the end of the Bible in Revelation, 18.13. Here, John lists various items of trade in a city that he calls Babylon. These articles include a number of perfumes and perfume ingredients: cinnamon, spice, incense, myrrh and frankincense. The ancient Chaldean city of Babylon was indeed famous for its trade in perfume materials. However, many parts of the Bible are intended to be metaphorical rather than literal and this passage is one of those. John is unlikely to be talking about that ancient imperial capital because it was razed to the ground many hundreds of years before John had his revelation on Patmos. When John refers to Babylon, he seems to be referring to Rome. In Revelation, 17.5, he describes Babylon as a woman, and in Revelation, 17.9, he says that the woman sits on seven hills. The city of Rome is built on seven hills. In John's day, to refer to Rome directly in the terms he uses in Revelation, 17, would have been very dangerous, hence I believe that he used metaphorical language which early Christians would have understood readily. However, whether he is speaking of Babylon or Rome, his message applies equally to the materialistic world which both cities typify and in which we still live.

The densest source of references to perfume is in the Song of Songs, sometimes known as the Song of Solomon (for example, Song of Songs, 1.3; 1.13; 3.6; 4.6; 4.10; 4.13; 4.14; 5.5; 5.13). This is a book of beautiful love poetry, sometimes expressing the thoughts of the lover, sometimes those of his beloved and occasionally interjections from their friends. The title "Song of Solomon" originated from the possibility that the lover in question was Solomon, David's son and successor as king and builder of the first temple in Jerusalem. Although being essentially romantic poetry, its place in the Bible can be justified on the basis of its being a metaphor for the love between God and humans.

1.3 METAPHORICAL AND LITERAL

Biblical references to perfume are sometimes metaphorical and sometimes literal and often are both simultaneously.

In the Song of Songs, the beloved tells her lover that his perfume is pleasing and so it is no surprise that all the girls love him (Song of Songs, 1.3), and she describes him as being perfumed with myrrh and incense (Song of Songs, 3.6). Such references are clearly intended literally. Similarly, when we are told that perfume and incense bring joy to the heart (Proverbs, 27.9), it is a simple literal statement.

Perhaps the best known reference to perfumes is found in the account of the visit of the magi to the infant Jesus (Matthew, 2.11), as prophesied by Isaiah (Isaiah, 60.6). The magi were probably Zoroastrian astrologers, and they brought gifts of gold, frankincense and myrrh. Since the magi came looking for a new king, it is not surprising that they brought expensive gifts for him. Thus, this account has a literal element; but it also has a metaphorical aspect. Two of the gifts are perfume ingredients and, as noted earlier, are included in both the first and last mentions of perfume in the Bible. Frankincense is an important ingredient in incense and hence has close associations with religious ritual and the priesthood. Myrrh contains pain killing substances and was a common component of mixtures used for embalming bodies. Consequently, it is associated with suffering and death. Thus, the three gifts brought by the magi each symbolise an aspect of Jesus' life. Gold represents his kingship, frankincense his priesthood and myrrh foreshadows his suffering in Gethsemane and death on Calvary.

The adoration of the magi and the gifts they brought to the infant Jesus have been portrayed in many ways by artists through the two millennia since their visit. Figure 1.1 shows a panel from one of the Bible Windows that can be found in the North Quire aisle of Canterbury Cathedral and it combines both the adoration of the magi and the visit of the shepherds. A different depiction in glass can be found in Troyes Cathedral, as shown in Figure 1.2. Le Mans Cathedral has a bas relief carving in wood showing the scene (Figure 1.3) and Chartres Cathedral has a scene comprising statuary in the local stone (Figure 1.4).

An example of a purely metaphorical reference to perfume can be found in the Apocrypha, where wisdom is described as being like a perfume containing cassia and spreading its fragrance like myrrh, galbanum, onycha and styrax (Ecclesiasticus, 24.13–17). In another example of metaphorical use of perfume, Saint John equates the prayers of the saints with incense in his Revelation (Revelation, 5.8).

Figure 1.1 Adoration of the magi.
Reproduced with permission from the Dean and Chapter of Canterbury.

Figure 1.2 Adoration of the magi depicted in a window of Troyes Cathedral.

Figure 1.3 Bas relief wood carving in Le Mans Cathedral showing the adoration of the magi.

One striking example of a description of perfume that is intended to be metaphorical but has deep literal significance can be found in Saint Paul's second letter to the Christian church in the Greek city of Corinth. Paul wrote, "But thanks be to God, who always leads us in triumphal procession in Christ and through us spreads everywhere the fragrance of the knowledge of him. For we are to God the aroma of Christ among those who are being saved and those who are perishing. To the one we are the smell of death; to the other, the fragrance of life." (2 Corinthians, 2.14–16, translation copyright © 1973 1978 1984 2011 by

Figure 1.4 Sculptures in Chartres Cathedral showing the adoration of the magi.

Figure 1.5 Saint Paul.
Reproduced with permission from the Dean and Chapter of Canterbury.

Biblica, Inc.™. Used by permission.) Here, Paul likens the spread of the Christian gospel through the world to the spreading of perfume through the air, and he also points out that different people respond very differently, both to smell and to the gospel. His analogy relies on two aspects of perfume and odour which he clearly understood well, though certainly not with the scientific detail that we have nowadays. This brings us on to the subject of how we perceive smell. Saint Paul is depicted in the South oculus of Canterbury Cathedral, as shown in Figure 1.5.

CHAPTER 2

How the Sense of Smell Works

2.1 THE ROLE OF SMELL AND TASTE

Smell and taste are the oldest of our senses. Both are based on the ability to detect and recognise chemicals present in the environment. Even the simplest single celled organisms have receptors on the cell surface that can detect chemicals in their environment. The basic mechanism then evolved in higher organisms to give the more sophisticated senses that we know as smell and taste. As far as humans are concerned, we have only five tastes: salt, sweet, sour, bitter and umami. Umami is the taste of glutamic acid and its salts such as monosodium glutamate. Glutamic acid is one of the 20 amino acids that are essential components of the proteins in our bodies, and this taste ensures that we consume sufficient protein in our diets. Similarly, the salt taste ensures that we consume a sufficient, but not excessive, quantity of electrolytes, and the sweet taste leads us to consume carbohydrates, a key source of energy for our bodies. The sour and bitter tastes give warnings of food that contains harmful bacteria and is beginning to decompose and also warn of toxins present in some plants, such as atropine in nightshade, and thus deter us from eating them. When we drink black, unsweetened coffee, the only taste we perceive is bitter; the remainder of the sensory experience is smell. Smell is a very complex sense, capable of detecting and discriminating between

Perfume in the Bible
By Charles Sell
© Charles Sell 2019
Published by the Royal Society of Chemistry, www.rsc.org

an essentially unlimited number of different molecules and combinations of these.

2.2 ANCIENT THEORIES OF OLFACTION

In Classical Greece, there was a philosophical argument about the composition of matter and how we perceive smells. Democritus (460–370 BC) argued that matter is made up of tiny indivisible particles that he called atoms (from the Greek ατομος (atomos) meaning indivisible) whereas Aristotle (384–322 BC) argued that matter is continuous and formed by blending, in different proportions, the four elements of fire, air, earth and water. In Democritus' theory, smell is explained by these tiny atoms entering the nose, and he postulated that pungent odours are caused by atoms with sharp edges that irritate the nose as they pass through whereas sweet odours are the result of atoms with smooth rounded surfaces. Democritus' concept of atoms travelling through the air is reflected in the word perfume, which is derived from the Latin *per fumum*, meaning through vapour. Smell is more difficult to explain using Aristotle's hypothesis, and he had to invoke the concept of emanations (radiation) from the source. Interestingly, in order to explain life, Aristotle invoked a fifth element (quintessence) or spirit. This led to the idea that, by extracting volatile substances from plants, we were removing the spirit, or life force, from the plant. Thus, distilled oils became known as "**quintessential oils**", nowadays shortened to "**essential oils**". Distilled alcoholic products, such as brandy, whiskey and eau de vie, are still referred to as "spirits", recalling Aristotle's theory. We now know that the odorous components of plants are discrete particles; not the simple atoms of Democritus' theory, but molecules made up from atoms of some of the 92 (not the 4 of Aristotelian cosmology) naturally occurring elements – in the case of odorants, mostly atoms of carbon, hydrogen, oxygen, nitrogen and sulfur.

2.3 THE MODERN UNDERSTANDING OF OLFACTION

The process involved in forming the **percept** that we know as odour is illustrated in Figure 2.1. It begins when receptor proteins in the nose interact with molecules in the air we breathe. The sensory cells where these receptor proteins are found are on the roof of the nasal cavity and so can only be reached by molecules that are

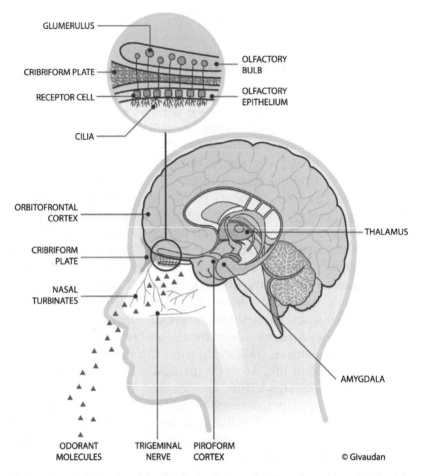

GLUMERULUS

CRIBRIFORM PLATE

RECEPTOR CELL

OLFACTORY BULB

OLFACTORY EPITHELIUM

CILIA

ORBITOFRONTAL CORTEX

THALAMUS

CRIBRIFORM PLATE

NASAL TURBINATES

AMYGDALA

ODORANT MOLECULES

TRIGEMINAL NERVE

PIROFORM CORTEX

© Givaudan

Figure 2.1 Schematic of the brain and nose showing the organs involved in the perception of odour.
Reproduced with permission from Givaudan.

sufficiently volatile. Odorous sources, such as flowers or freshly baked bread, contain odour molecules that vaporise into the air around them and are then carried in the air to the receptors in the nose. Hence, Saint Paul is using this simple scientific observation when he talks about fragrance spreading (2 Corinthians, 2.14–16). Similar references to fragrance spreading can be found in the Song of Songs where, for example, the beloved talks about her perfume spreading across the table to reach the king (Song of Songs, 1.12) or when she invokes the wind to spread the fragrance of her garden to attract her lover to it (Song of Songs, 4.16). Similarly, we read that the mandrakes send out their fragrance (Song of Songs, 7.13) and the vine blossoms spread theirs (Song of Songs, 2.13). When someone wearing a perfume walks through a room, they leave a scent trail in the air, known in perfumery as silage, and this is another example of perfume spreading from the wearer to the air around them.

The nose has traditionally been seen as the organ responsible for the phenomenon of smell and this is evident also in the Bible. In the Old Testament, Psalm 115.6 says of carved idols that "they have noses but cannot smell" and in the New Testament, Saint Paul asks "If the whole body were an ear, where would the sense of smell be?" (1 Corinthians, 12.17). The nose is certainly important since it is in the nasal cavity that molecules from the environment are detected, but odour exists only in the brain and so we smell with our brains, based partly on signals from the outside world that are first detected in the nose.

2.3.1 Nasal Metabolism

Recent research has shown that the nose plays another role, and one that the writers of the Bible would not have imagined. The mucus in our noses is rich in a variety of **enzymes** that carry out chemical reactions on substances that we inhale. This serves a number of important roles, but it also means that one odorant molecule might be transformed by one or more of those enzymes into another molecule with a different odour. One example of this is the transformation of the woody odorant, shown in Figure 2.2, by the enzyme CYP2A13 into a molecule that elicits a raspberry odour.[1-5] Some people have this enzyme in their nasal mucus while others do not. Those with the enzyme detect a raspberry odour, and those without the enzyme smell wood. Thus, individuality of perception begins even before the odour receptors in the nose have been activated.

2.3.2 Recognition of Odorants and Synthesis of an Odour Percept

Humans each have about 10 million sensory cells in the nose to detect odorous molecules. Each one has 10–15 hair-like

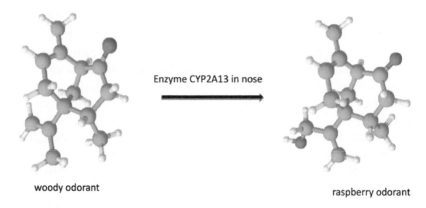

woody odorant

Enzyme CYP2A13 in nose

raspberry odorant

Figure 2.2 Nasal chemistry.

projections called cilia that dangle into the mucus on the top surface of the nasal cavity. The receptor proteins that detect and recognise odour molecules are located on the surface of these cilia and each sensory cell contains only one type of receptor protein. We each use about 350–400 different types of receptor protein and there may be several variants of each type. Each type of receptor protein responds to a range of different molecules. Some finely tuned receptors are selective as to which molecules they detect, whilst others are very broadly tuned and respond to a wide variety of different molecules. Conversely, each type of odour molecule is capable of stimulating a number of different types of receptor protein. Thus, there is no simple code, and it is the overall pattern of activation that is important. Research in the last few decades into the odour receptors and other receptors of the same type has led to the award of several Nobel Prizes.

The receptor proteins adopt a shape consisting of seven helical columns assembled into one larger column. This large column sits across the cell membrane and has a pocket on the outer surface, between the seven helices, into which odour molecules can fit. It is the precision of the fit between the two that constitutes recognition of the odour molecule by the receptor. When a suitable odorant enters the pocket on a receptor protein, the pocket closes around it. In doing so, the protein molecule adapts its shape and a different pocket opens up on that part of the receptor protein that is inside the cell membrane. Another protein, called a G-protein, then binds into this new pocket and a chain of chemical reactions is set in motion inside the cell. These sensory cells are nerve cells (neurons) with one end dangling into the nasal cavity and the other end in the olfactory bulb, the first part of the brain that is involved in processing smell signals. Thus, the changes induced by recognition of an odour molecule in the nasal cavity result in generation of an electrical signal in the brain. Because each type of molecule can stimulate a variety of sensory neurons and each sensory neuron can be stimulated by a variety of different molecules, when a mixture of different chemicals (which all natural scents are) enters the nose, a complex pattern of electrical signals is received by the olfactory bulb. This signal pattern is sent to a brain region known as the piriform cortex where it is processed further and electrical signal patterns known as odour objects are formed. From the piriform cortex, signals spread out through various regions of the brain where they interact with input from other senses and from cognitive factors, such as experience, context and expectation. These various signals then come together in a brain region known as the orbitofrontal cortex, and there, the percept of odour is synthesised by the brain. Odour has no physical reality; rather, it is the effect of an electrical stimulus in the conscious part of the brain. An example of interaction with other senses is the fact that wine experts are unable to correctly describe the flavour of a white wine if it contains an odourless, tasteless red dye. The visual input of the red colour distorts the way in which the odour signals are interpreted by the brain, and so the odour percept that is formed in the orbitofrontal cortex is not the same as it would be if there was no red dye in the wine. Similarly, interpretation of the odour object is affected by the context in which the experience occurs, by the expectations of the person doing the smelling and by their previous experience of smelling different odours.

2.3.3 The Individuality of Perception

We know that the variation in the array of receptor proteins from one individual to another is so great that it is unlikely that any two humans, with the possible exception of identical twins, use the same pattern of receptors in the nose. Differences in brain

pathways, experience and so on then add further levels of di-
versity, and so any two people smelling the same odour source
will have different odour percepts in their orbitofrontal cortices.
Since we learn to give the same names to smells, we are then
tricked into thinking that we perceive them in the same way as
the other person. So, for example, two people smelling a rose will
both say that the odour is that of rose and will think that they
each have the same mental percept, whereas the reality is that
they each have a unique percept and neither will ever know
what the other really perceives. We each live in our own unique
sensory universe. Furthermore, whether someone describes
any given scent as pleasant or unpleasant will depend on that
person's individual experience of life. For example, someone
who associates the odour of cloves with apple pie will react
positively to the odour, unlike someone who associates the
odour with a distressing episode at the dental surgery. The more
deeply I look into the mechanism of odour perception, the more
I agree with the psalmist who wrote that "we are fearfully and
wonderfully made" (Psalm 139.14).

So, when in his letter to the church in Corinth (2 Corinthians,
2.14–16), Saint Paul wrote about the different response to the
gospel as being like the different responses to the same odour,
he was touching on a profound truth about human individuality
that has only been fully realised through modern scientific
research.

The French philosopher René Descartes (1596–1650) said
"I think therefore I am", and this is the only certainty that any of
us can have. Our understanding of the universe we live in is a
mental model built up from input from our senses. The hymn
writer F. S. Pierpoint (1835–1917) hinted at this when he wrote
the following verse. In this context, he uses the word sense
to mean understanding, and so it is our understanding of the
information coming in to our brains that enables us to create a
mental model of the universe.

For the joy of ear and eye,
For the heart and brain's delight,
For the mystic harmony
Linking sense to sound and sight,

Pierpoint's contemporary, the psychologist William James, was more specific when he wrote in his book *Principles of Psychology* (1890), "Whilst part of what we perceive comes through our senses from the object before us, another part (and it may be the greater part) comes from inside our heads". So, at least in the case of smell but probably also for the other senses, we each live with a unique mental model of the universe around us because each of us receives different signals from our sensory organs and each individual brain constructs mental models in its own way.

2.4   BODY ODOUR

Just as we each have a unique sensory experience of the world around us, we each have a unique body odour. Modern research has shown how human body odour is produced by the action of bacteria on the odourless chemicals that we humans produce and excrete/secrete onto the skin. The bacteria live on the substances secreted by the human and, in doing so, they release small odorous molecules. The chemicals secreted by the human differ in different parts of the body, and the chemistry by which they are converted to odorous chemicals depends on the nature of the bacteria involved. Different species of bacteria tend to be found in different parts of the body: *Staphylococcus* and *Corynebacterium* species in the armpits and *Pittosporum ovale* on the scalp, for instance. Different individuals produce different combinations of skin chemicals, and different individuals have different mixtures of bacteria on the skin. Hence, each individual human has a distinctive odour, and this is how bloodhounds can track one person's scent trail even when it crosses those of other people. It is also how Isaac, who had gone blind, was deceived by his son Jacob into thinking that it was Jacob's brother Esau he was talking with when Jacob wore his brother's clothes (Genesis, 27.27). Esau's distinctive body odour would have been adsorbed onto his clothes, and since Isaac was blind, he would have relied on smell, sound and touch to know who was with him. Jacob's wife Rebekah had coached Jacob into this deception using goat skin to give the appearance of hairy skin like Esau's rather than Jacob's smooth skin, and she had also prepared a meal for Isaac in the way that Jacob did in order to deceive Isaac's sense of taste. Figure 2.3 is from the Ancestors series of windows in Canterbury Cathedral and depicts Isaac.

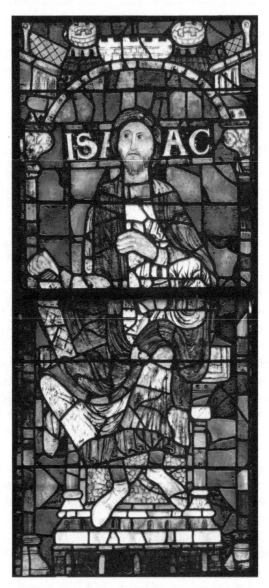

Figure 2.3 Isaac.
Reproduced with permission from the Dean and Chapter of Canterbury.

2.5 FURTHER READING

For a more detailed and technical account of this subject and for evidence supporting the various statements made here, see *Chemistry and the Sense of Smell*, *Learning to Smell* and *Neurogastronomy*. Details of these books can be found in the bibliography.

REFERENCES

1. B. Schilling, R. Kaiser, A. Natsch and M. Gautschi, *Chemoecology*, 2010, **20**, 135–147 and references cited therein.
2. CH2005/000412, B. Schilling, to Givaudan.
3. WO 2006/007751, B. Schilling, to Givaudan.
4. WO 2006/007752, B. Schilling, to Givaudan.
5. WO 2008 116338, B. Schilling, T. Granier, G. Fráter, A. Hanhart, to Givaudan.

CHAPTER 3

Perfume Ingredients in Nature

3.1 BIOSYNTHESIS

Whereas animals obtain the chemicals they need for life by eating plants or other animals, plants make the vast array of necessary molecules from carbon dioxide and nitrogen that they extract from the air, water and inorganic chemicals that they obtain from the soil that they grow in. Sunlight provides the energy that they need to carry out the chemical processes used to build complex molecules from these simple starting materials. Carbohydrates, proteins, fats and nucleic acids are chemicals that are essential for all living organisms and so are referred to as primary metabolites. In contrast, secondary metabolites are those that are produced by some organisms and not others. The volatile chemicals responsible for the scents of plants, and hence those used in perfumes, are secondary metabolites. The chemical processes used by plants to make these follow a number of defined pathways and each pathway produces molecules with distinctive features of molecular structure. One of the key pathways builds different molecules using an intermediate substance called **shikimic** acid and chemicals produced from this pathway are known as **shikimates**. The principal odour molecules of cinnamon and cassia belong to the shikimate family of chemicals. Another major pathway uses an intermediate known as isopentenyl pyrophosphate and this gives a family of chemicals

Perfume in the Bible
By Charles Sell
© Charles Sell 2019
Published by the Royal Society of Chemistry, www.rsc.org

known as **terpenoids**. Terpenoids are characterised by having molecular structures containing multiples of five carbon atoms arranged in a specific pattern. My specialisation is in terpenoid chemistry and so I am pleased to see that most of the Biblical perfume ingredients – such as frankincense, myrrh, calamus and spikenard – owe their odours to terpenoid molecules. Terpenoids display a vast assortment of intricate molecular structures and so I can appreciate the ingredients' beauty not only in terms of odour but also in terms of the wonderful molecular architecture of their components.

3.2 THE ROLE OF PERFUME INGREDIENTS IN NATURE

The secondary metabolites produced by plants vary from one plant to another, giving each species a distinctive chemistry. The chemical compounds are produced for specific purposes that are important for the species concerned. As far as the plants used as perfume ingredients in the Bible are concerned, the commonest reason for synthesis of the odorant molecules is for defence.

When the bark of a tree is damaged, the tree is vulnerable to attack by bacteria and fungi entering the wound. A common response to this is for the tree to produce a liquid in the wound that contains chemically reactive small molecules that can link together (**polymerise**) when exposed to air. The liquid thus hardens into a solid substance that seals the wound and prevents bacteria and fungi from entering. Typical examples are the formation of pine rosin in pines, firs and spruces and rubber in the rubber plant. In addition to this physical barrier to infection, many trees also produce antibacterial and antifungal compounds that are incorporated into the solid, providing a level of chemical defence in addition to the physical defence. Some of the antibacterial and antifungal compounds are volatile and these are often then responsible for the odour. Examples of such odour containing resins are frankincense and myrrh.

Plants can also produce antibacterial and antifungal compounds in their leaves, bark, roots and rhizomes before injury in order to provide a ready chemical defence if the plant comes under attack from microorganisms, and these are described as physiological products, unlike those described in the preceding paragraph, which are described as pathological products.

The defensive chemicals of cinnamon, cassia, galbanum and labdanum are examples of such physiological products. The underground parts of plants, such as roots and rhizomes, are particularly vulnerable to microbial attack and so these often contain antibacterial and antifungal substances. Underground parts need to be protected against insects, worms and the like as well as against bacteria and fungi. Therefore, rhizomes often contain an array of "chemical weapons" against such attack. Examples where this gives rise to odorous molecules in rhizomes are spikenard and calamus.

When plant material undergoes fungal attack, in addition to defensive chemicals produced by the plant, the fungi produce antibacterial compounds to protect themselves from bacteria, and it is these processes that lead to the odour molecules present in agarwood.

Once a substance has been produced by a plant, it becomes subject to decomposition by various different processes, and decay can result in formation of smaller, volatile molecules from larger ones. This is also a source of odorants and the scents of saffron and orris are the result of decay and degradation of molecules produced by the plant.

3.3 FURTHER READING

More detailed accounts can be found in *The Chemistry of Fragrance, Chemistry and the Sense of Smell, The Chemistry of Essential Oils, Handbook of Essential Oils* and *Figs, Dates, Laurel and Myrrh* details of all of which can be found in the bibliography. The classical books on **essential oils** are both now out of print but if copies can be found through a library, they are well worth studying. Both *Die Ætherischen Öle*, and *The Essential Oils* are authoritative multi-volume books.

CHAPTER 4

Sources of Perfume Ingredients

Of all the perfume ingredients mentioned in the Bible, only galbanum is a true native to Palestine. Several of the ingredients come from the Horn of Africa and the southern tip of the Arabian Peninsula. Others come from much further afield. Labdanum grows across the Mediterranean region but the main source is Spain. Similarly, orris is produced mainly in Italy. Cinnamon is produced in Sri Lanka, cassia in China and agarwood in China and Vietnam. Thus, even in early Biblical times, trade routes operated across Europe and Asia. Ezekiel mentions Danite and Greek merchants trading cassia and calamus through the port of Tyre (Ezekiel, 27.19). We know that the Phoenicians and Greeks had extensive trade routes right across the Mediterranean, stretching even to Britain (tin and gold) and Denmark (amber). Similarly, the trade route later known as "The Silk Road" stretched from the Middle East across to China. So, it is not surprising that the perfume ingredients of the Bible came from as far west as Spain, as far east as China, as far south as Sri Lanka and from the tops of the Himalayas to the bottom of the sea.

4.1 EXTRACTION OF PERFUME INGREDIENTS

Nowadays, odorous components are usually extracted from plant material by solvent extraction, distillation or a combination of the

Perfume in the Bible
By Charles Sell
© Charles Sell 2019
Published by the Royal Society of Chemistry, www.rsc.org

two. A fuller account of modern methods of extraction of perfume ingredients from plant sources can be found in Chapter 3 of *The Chemistry of Fragrance*.

Distillation is the most important modern technique and is usually carried out in the presence of water. A simple schematic drawing of such a still is shown in Figure 4.1. The steam generated on heating water in the still pot helps to release the volatile odorants from the plant matrix and carry them in the vapour phase to a cooler surface where the steam condenses to water and the plant oils also condense to liquid. The liquefied materials then drain down into a receiver. Since the plant oils are insoluble in water, the two separate into different layers (known to chemists as phases). The water (or aqueous) layer is usually the heavier of the two and can be run off through a tap at the bottom of the receiver, and then, subsequently, the oil can be drained off. In practice, a device known as a Florentine flask, or an equivalent apparatus, is used, and this allows continual removal of the water from a drain at the bottom of the receiver and removal of the oil from an outlet higher up. The water is usually recycled to the still pot. Commercial scale distillation of rose oil is illustrated in Figures 4.2 and 4.3. Figure 4.2 shows bags of rose petals ready to be loaded into a still, and Figure 4.3 shows the petals after charging to the still.

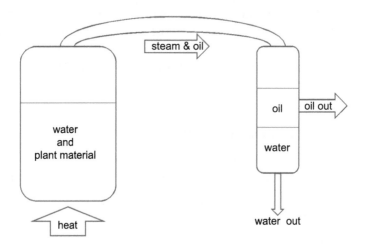

Figure 4.1 Schematic diagram of a still and Florentine flask.

Figure 4.2 Rose petals ready to be charged to a commercial still.
Reproduced with permission from Givaudan.

Figure 4.3 Rose petals in a commercial still.
Reproduced with permission from Givaudan.

Modern solvent extraction usually employs a solvent such as hexane or a mixture of hexane and ethyl acetate. Treatment of the plant material with the solvent results in removal of the plant oils into the solvent. The solvent is selected as one with a much lower boiling point than the odorous plant oils, and so it can be removed from the plant oils by distillation. Solvent extraction removes other substances along with the essential oils of the plant. These other substances would include **lipids** (fatty oils also known as fixed oils) and oil soluble materials such as chlorophyll, the green pigment present in leaves. The material left after removal of the extraction solvent is known as a **concrete** and usually has the appearance of a fatty solid. The odorous oil can be removed from the concrete either by distillation, since the lipids *etc.* are much less volatile, or by extraction with ethanol (alcohol), since the lipids (fats) are much less soluble in ethanol, especially if the ethanol contains a small amount of water.

The only liquid solvent available in Biblical times was olive oil, and so fragrant oils could be made by treating the plant material with olive oil and filtering off the plant fibre and other insoluble substances. Of course, the perfume ingredients could not be separated from the oil and would have to be used as a solution in the oil. Another method that was used in Biblical times was one that became known more recently as "**enfleurage**", a method that is still in use. In this technique, the plant material is mixed with animal fat and, once the perfume oils have migrated from the plant leaves, flowers or whatever into the fat, the fat is melted and filtered to remove fibres and other solid material and then resolidified to give a product also known as a concrete. Nowadays, the concrete is usually extracted with ethanol to separate the odorous oil (which is soluble in ethanol) from the animal fat and the plant fats, waxes and other ethanol insoluble substances. This would not have been possible in Biblical times and so the concrete would have been used as such and would have looked more like an ointment. Egyptian tomb paintings often show people with cones of concrete (perfumed fat) on their heads. In the hot climate of Egypt, the fat would melt and the perfumed molten fat would run down through the hair and onto the body of the wearer. An Egyptian harpist's song from about 1400 BC advises listeners to "Put perfume on your head and clothe yourself with fine linen". In Figure 4.4, a wall painting from the tomb of Nebamun (*ca.* 1350 BC) shows a group of women at a banquet, each of them wearing a cone of perfumed fat on her head.

Figure 4.4 Wall painting in the tomb of Nebamun.

The use of olive oil as a solvent for extraction or handling of perfumes and the technique of enfleurage into fat introduce an issue that will be discussed also in Chapter 9. In that chapter, it is the Greek word μυρος (muros) that is of interest. Here, in the Hebrew Bible, it is שֶׁמֶן (še·men), שְׁמָנִים (šə·mā·nîm) and related words that raise the same questions. Hebrew uses idea associations leading to one word having different meanings in different contexts. So, שֶׁמֶן (še·men) and שְׁמָנִים (šə·mā·nîm) in most cases are translated as oil but are also used to mean, for example, a medical ointment, fertility, transparency, joy, hospitality, wild olive trees and luxury. In some instances, the context suggests or implies a perfumed oil or concrete. When it is qualified as sacred anointing oil, then it is almost certainly a perfumed oil. In other cases, for instance in the preparation of food, it is unlikely to mean a perfumed oil. Between the two extremes, there are cases where the oil is added to grain or bread for burnt offerings or used as fuel for lamps. In those examples, the oil might, or might not, have been perfumed.

A third technique available in Biblical times and still in use today involves boiling the plant material in water. The perfume components, along with oils, fats, waxes, gums and resins are separated from the plant material by the boiling water and, when cooled, the organic extracts separate from the water, usually as a sticky gum or resin. This method is still used, for example, in the extraction of labdanum. In fact, the Greek word (in both Greek of Biblical times and today) for perfumer is μυρεψικος (murepsikos), which comes from the verb μυρεψεω (murepseo), which means I boil or prepare unguents and which also gives the word μυρεψειον (murepseion), meaning perfume factory, thus supporting the view that this was a major method of extracting perfume ingredients from plant material. The gum or resin extracted in this way could either be used as such or dissolved in fat or an oil such as olive oil. Many of the ingredients discussed in Chapter 6 would have been prepared in this way by boiling the raw material, labdanum and onycha, for example. The plant material left after extraction of the perfume ingredient could be dried and used as fuel to boil the water for the next batch. In the same way, field stills can be fuelled by spent plant material from the previous batch. This practice was used in some countries in the 20th century and might continue to be used today.

Colognes, eaux de parfum and other perfumes for application to the skin – in other words, the products that spring immediately to the 21st century mind when thinking of perfume – are solutions in ethanol–water mixtures, and so we are accustomed to thinking of perfume as a free-flowing liquid. Distillation was invented by the Arabs in about the year 1000 AD, and so was unknown in Biblical times. Ethanol is produced by distillation from fermented fruits and vegetables, and so it was also unknown as a pure liquid in Biblical times. So, Biblical perfumes looked very different from what we are used to nowadays.

In the case of incense, the odorous chemicals were released into the air from the plant material by heating; in fact, by burning. In such use, then, the term *per fumum* took on a very literal meaning of "through smoke". Perfume materials are susceptible to degradation by the action of bacteria and other microorganisms, sunlight and oxygen in the air. The use of edible oils and fats to extract the odour molecules from the plant material would make the perfume even more prone to degradation by bacteria, especially in Biblical times when the fat or oil would also serve as the carrier. The book of Ecclesiastes reminds us that dead flies give perfume a bad smell (Ecclesiastes, 10.1). This would be through the action of bacteria carried into the perfumed fat by the flies. One of the plagues to hit Egypt in the days of Moses affected the River Nile and all the fish died. The river developed a foul smell, presumably because of decomposition of the dead fish (Exodus, 7.21). Similarly, in the Sinai desert after their escape from Egypt when the Israelites were given manna to eat, if they disobeyed Moses and kept the manna for too long, it became infested with maggots and developed a bad smell (Exodus, 16.20). Once again, bacterial action was probably responsible for converting food into volatile malodorous substances.

4.2 STORAGE OF PERFUMES

Air, heat and light are all capable of causing degradation of perfumes, especially when all three are present together. So, the best place to keep perfume is in a full, well-sealed bottle in a refrigerator, something that was impossible in Biblical times. Perfume for personal application is nowadays mostly kept in clear glass bottles, usually **mould blown** and usually with a pump spray device in the stopper. Perfume bottles were rather different in Biblical times. Excavation of sites used in ancient times as perfume factories have provided samples of perfume bottles made of earthenware, stone and glass. Perfumes were very expensive in Biblical times, owing to the cost of the raw materials and the labour involved in preparing perfume from them. Similarly, the bottles containing them were also very expensive for similar reasons but with labour costs being even more pre-dominant. In the eighth century BC, Isaiah included perfume bottles along with fine clothes, jewellery, purses and mirrors as the trappings of wealth, luxury and decadence (Isaiah, 3.18–20).

Earthenware bottles were made by forming clay into the desired bottle shape and then baking it in a simple oven such as a kiln or pit. Nowadays, the oven temperature would be about 1000 °C, but early ovens might have achieved only 600 °C. The colour of the bottle would vary according to the exact com-position of the clay. Higher iron content would result in the typical reddish colour of terracotta. The simple fired earth con-tainer would be porous but glazes could be used to make it water, and perfume, tight. The bottles were often painted before glazing. Figure 4.5 shows an ornately shaped Greek earthenware perfume bottle, and one that is decorated with a painting of an Amazon is shown in Figure 4.6. Inspection of the top of this bottle, and those of the bottles shown in Figures 4.5 and 4.7, makes it clear why stoppers were not as well-fitting in the ancient world as are those of today that have plastic linings to ensure a tight seal, as mentioned below.

Bottles could also be carved out of relatively soft stone and a preferred material was alabaster. There are two different minerals that go under the name of alabaster. The harder one is gypsum alabaster, which is a sulfate of calcium, whereas the softer calcite alabaster is a carbonate of calcium. Calcite

Figure 4.5   Perfume bottle in the shape of a woman holding a dove, made on Rhodes 530–520 BC.
Copyright © The Trustees of the British Museum. All rights reserved.

Figure 4.6 Perfume bottle made in Athens about *ca.* 470 BC.
Copyright © Universal History Archive/Universal Images Group/Getty Images.

alabaster is the one that was used in the Middle East in Biblical times.

The Greek for alabaster is αλαβαστος (alabastos), and alabaster was so prominent as a material for manufacture of perfume bottles that the Greek word for perfume bottle is αλαβαστρον (alabastron) and it is used even for bottles made of earthenware or glass. The Greek word is thought to be derived from the town of Alabastron in Egypt where the stone is quarried. That name, in turn, could be derived from the fact that alabaster vessels were used in the cult of the Egyptian goddess Bast.

An alabaster perfume bottle was found in the temple of the Sumerian goddess Inanna in southern Iraq and has been dated to 3200–3000 BC. One has been found that was made on the Greek island of Rhodes about 1750 BC, and one was found in the

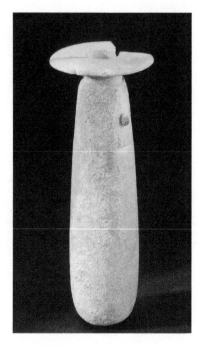

Figure 4.7   Perfume bottle made of alabaster.
Copyright © Science & Society Picture Library/Getty Images.

tomb of Tutankhamun dating from 1323 BC. The illustration in
Figure 4.7 shows a typical example of an ancient perfume bottle
made from alabaster.

Glass was discovered about 3500 BC, possibly in Syria, Egypt or
Mesopotamia. Initially, glass was used as a glaze for earthenware
vessels, and Egyptian examples date from the 12th dynasty
(before 3500 BC). The oldest Egyptian glass bottles appeared
later, in the 18th dynasty, *ca.* 1500 BC, and were used to store
perfume and **kohl**. Glass is made by heating sand with an alkali.
In Egypt, this might have been **natron** from the deposits around
the Faiyum area just south of the Nile delta. The ash produced by
burning certain plants, particularly seashore plants, is another
source of sodium carbonate and would have been particularly
important for Syrian glass makers. Glass making in Egypt went
into decline between 1100 and 600 BC then was regenerated,
possibly because the port of Alexandria offered an opportunity
for Mediterranean trade in competition with the glass makers of
Syria. Almost two millennia later, Venice became a major source

of glass manufacture, and it is possible that the Venetians acquired the skills and trade secrets from their 12th century trading posts in Syria, and Tyre in particular, or by capturing craftsmen when they took Constantinople in 1204 AD. Medieval glass makers in Northern Europe used wood ash, which is richer in potassium carbonate and gives a different quality of glass. Natron and both of the ashes are alkaline and disrupt the chemical structure of the silica in the sand to form glass that has a lower melting point than silica. Addition of calcium, for example as lime, helps to prevent the glass from dissolving in water and increases its resistance to corrosion. Like perfumers, early glass makers kept their trade secrets very closely guarded, and glass workers had to import glass ingots from those who held the secret technology. The earliest glass was coloured and opaque owing to impurities in it. Technology for making clear glass appeared around the ninth century BC. The colours are due to metal ions trapped in the glass matrix, and even clear glass tends to have a greenish tint because of iron dissolved in the glass. Iron is ubiquitous and it is difficult to produce glass free from it. If a pane of modern window glass is viewed end on, this green hue is almost always in evidence. These metals enter the glass initially as oxides or salts, and much research went into controlling the colour by choice of additives. In the early 14th century AD, silver stain was discovered when it was found that application of silver oxide to its surface resulted in the silver migrating into the glass and staining it yellow.

The earliest technique for making glass bottles is known as **core forming**. In this technique, a core was made from dung, clay, sand and water, placed on a metal rod, formed into the desired interior shape of the bottle and dried. The mould was then covered with a layer of glass. To decorate the bottle, strings of molten glass were wound around the core and the use of different colours of glass would result in patterns on the final bottle. While the glass was still soft, a blade could be used to add further artistic effects, as was done with the bottle shown in Figure 4.8. If handles were required, they were added separately using threads of molten glass. The whole assembly was then heated to **anneal** the glass; in other words, to cause it to meld together into one piece. After cooling, the metal rod could be removed from the core by tapping it sharply and the core was

Figure 4.8   A core formed glass bottle.
Copyright © Mondadori Portfolio/Hulton Fine Art Collection/
Getty Images.

scraped out to leave the finished bottle. Crete was an important
centre of glass manufacture, and bottles in what is known as
Greek style core formed vessels appeared from about 600 BC.
Core formed bottles tended to be slender and often measured
only up to 10 cm or so in length. A typical example is shown in
Figure 4.8.

During the first millennium BC, glass bottles were also made
by casting in moulds. The desired shape of bottle was made in
wax and this was covered with clay or plaster. Firing the mould
would cause the wax to melt and run out and it was replaced by
sufficient molten glass to coat the interior of the mould but leave
an empty interior. This technique was used alongside core
forming until the invention of blown glass.

Around the first century BC, Phoenician glass workers living
on the coast of Syria or Judea discovered that, if a blob of molten
glass was attached to one end of a metal tube, the glass worker
could blow down the other end of the tube and create a hollow

glass bubble that could be fashioned into a bottle while the glass was still soft. Thus, the skill of glass blowing was born. Figure 4.9 shows some examples of blown glass bottles made during the days of the Roman Empire. These bottles are in the Gallo-Roman museum of Sens so probably are of later date than the first century AD but they do illustrate the type of bottle that was becoming available in New Testament times. Two of them have handles that would have been added after forming the bottle itself, by applying the appropriate piece of molten glass to the bottle then annealing the whole to ensure a robust joint. These bottles show the greenish tint due to iron traces in the glass, as was mentioned earlier in this chapter. Towards the end of the

Figure 4.9 Blown glass bottles.

first century AD, the technique of blowing glass into a mould was developed, giving rise to what is known as mould blown glass and opening the way for larger scale production of glass bottles. Blown glass was cheaper than pottery or carved stone, and consequently, glass bottles became the commonest vessels for perfume. Using terracotta moulds, very intricate designs of bottles could be made. Production moved north from the Mediterranean (Syria, Cyprus and Crete) and Cologne became the major glass blowing centre of the Roman Empire. Modern glass perfume bottles are mould blown.

Stoppers were made of earthenware, cork or wood and did not form such a good seal as we are used to nowadays since flexible plastic seals were not available. The bottle necks were also not so cleanly formed as those of today and this also prevented a tight seal from the stopper. Therefore, in order to protect the contents from ingress of air and from leakage, the stopper was sealed in place with wax or some other suitable material, such as fine leather. In order to open the bottle, the seal would have to be broken. The British Museum collection includes two blown glass bottles made in the second half of the first century AD (therefore possibly in late New Testament times) that were made in the shape of birds. These bottles were filled with perfume or cosmetics and then sealed by melting the glass around the opening through which they had been filled. It was then necessary to break the birds' tails in order to remove the contents.

CHAPTER 5

Identifying Perfume Ingredients in the Bible

5.1 LANGUAGE

The word perfume is sometimes used to describe what a perfumer would recognise as an ingredient or component of a perfume. Similarly, some words might mean either an ingredient or a compounded perfume. For example, the word rose might mean an extract of rose, such as rose otto, or it might mean a compounded perfume with a distinctly rose like character. To use a Biblical example, the word incense should refer only to a compounded perfume whereas frankincense refers to a specific ingredient used in incense and also in other perfumes. The Hebrew Bible does make a clear distinction between incense and frankincense. When frankincense is intended, the Hebrew Bible uses the word לְבֹנָה (lə·ḇō·nāh), which is translated as λιβανος (lib'-an-os) in the Septuagint and other Greek translations and as frankincense in English, although some translations tend to use the word incense when the Hebrew states לְבֹנָה (lə·ḇō·nāh), *i.e.* frankincense (for example, in Numbers, 5.15). Incense, on the other hand, is a mixture of different ingredients, one of which could be frankincense, and the Hebrew uses קְטֵר (qiṭ·ṭēr) or קְטֹרֶת (qᵉṭō·rĕṭ) to describe it. In some instances,

Perfume in the Bible
By Charles Sell
© Charles Sell 2019
Published by the Royal Society of Chemistry, www.rsc.org

such as Exodus, 30.1, the Hebrew adds the word מִקְטָר (*miq·ṭār*), which means burning, indicating that the incense is intended to be burnt on an altar or in a censer. In Ezekiel, 23.41, the word is modified to וּקְטָרְתֶּי (*ū·qə·ṭā·rə·tî*), which means my incense, and since it is God speaking through the prophet, this probably refers to the holy incense of Exodus, 30.34–35. In some other cases, we cannot be sure which incense formulation is used.

Smell is unlike our other senses (except for bitter and sweet tastes) in that it has no fixed physical reference points. Vision can be measured through intensity and wavelength of light, hearing by frequency of vibrations in the air, touch by pressure, salt taste by chloride ion concentration and sour taste by pH (degree of acidity). Odour descriptions are always associative; in other words, we describe a smell as being like another smell. When a perfumer or other expert in perfumery describes a perfume, they do so in terms of the ingredients it contains. The odours of ingredients are usually classified roughly according to volatility. Thus, top notes (such as citrus and pine) are the most volatile ones and the first that we perceive on opening a bottle of perfume. Middle notes (such as rose or jasmine) represent the central core of a modern fragrance, and the end notes (such as musk and vanilla) are those that are last to disappear from the perfumer's blotter, skin, fabric or whatever substance the perfume is applied to. The terms top, middle and end notes are often replaced by the more romantic sounding head, heart and base. Taking the perfume "Angel" as an appropriate example of a modern fragrance for a book on Biblical perfumes, it might be described as having a top note of bergamot and dewberry with a floral, rose heart on a base of patchouli, chocolate and vanilla.

5.2   CONTEXT

With this in mind, when I searched for words in the Bible, I used terms such as rose, lily, frankincense and spikenard to describe ingredients as well as words like perfume, scent, incense and odour that might be used to describe perfumes. The searches were carried out in English, Hebrew and Greek. The results were somewhat surprising since many materials which are used as perfume ingredients nowadays are mentioned in the Bible but not in the context of perfumes. Table 3.2 of *The Chemistry of Fragrances* lists 84 plant oils and extracts in common use nowadays whereas I found only a dozen or so in Biblical texts.

Roses and lilies are mentioned but always as flowers and not as ingredients of perfume. In the Song of Songs, the beloved describes herself as being like a rose of Sharon or a lily of the valleys (Song of Songs, 2.1) and she describes her lover's lips as being like lilies (Song of Songs, 5.13). In the Sermon on the Mount (Matthew, 6.28), Jesus compares the lilies of the field to fine clothes but does not comment on their scent, let alone any use in perfume.

Similarly, cedarwood and sandalwood are both significant ingredients in many modern perfumes but, although both woods are mentioned in the Bible, it is always in the context of building materials rather than as perfumes. Cedarwood is אֲרָזִים ('ă·rā·zîm) in Hebrew {ξύλον κέδρινον (ksulon kedrinon) in the Septuagint} and features strongly both in the building of the tabernacle in the desert (*e.g.* Leviticus, 14.4 and the rest of that chapter) and the building of Solomon's temple in Jerusalem (*e.g.* 1 Kings, 5.6 and subsequent chapters). The Hebrew word אַלְמֻגִּים ('al·mug·gîm) is usually rendered as almug or algum wood in English trans-lations, as for example, in 1 Kings, 10.11, where it is described as a building material for Solomon's temple and the wood used in making harps and lyres. Sandalwood, juniper and box have all been suggested as interpretations, although a different Hebrew word (וּתְאַשּׁוּר ū·tə·'aš·šūr) is used for box in Isaiah, 41.19 and 60.13. Unlike other possible candidate species, sandalwood is not native to Israel but the context of the references in Kings and Chronicles do not indicate whether the wood is local or imported. Modern Israeli scientists identify it as sandalwood.[1] However, the Septuagint (III Kings, 10.11 – equivalent to 1 Kings, 10.11 in the English Bible) translates it as ξύλα πελεκητά (ksula peleketa), meaning hewn timber. Other woods – citron wood, terebinth, myrtle and pine – are also mentioned as building materials, and so the tabernacle in the desert, David's palace and Solomon's temple must all have had wonderful woody scents in the air inside them. It is not so surprising that the woody notes of modern perfumery are absent from Biblical perfumes. Nowadays, the essential oils of woods are extracted by cutting the wood into small chips (usually shavings, sawdust and scrap from carpentry) and then steam distilling these to produce the oil. The addition of water to the distillation helps to remove the oils from the wood and vaporise them and to prevent charring or **pyrolysis** of the wood and its oil. Distillation was not invented until about 1000 AD and so would not have been available in Biblical times. However, some woody odours, such as camphor, can be extracted by boiling the wood in water and so might have been available in Biblical times.

There are many mentions of herbs in the Bible, all of which are used in modern perfumes but are discussed only in other contexts in the Scriptures. Coriander, cumin, dill, wormwood, hyssop, mint and rue are all included. Mint and rue are mentioned specifically as tithes (Luke, 11.42) and hyssop is used for purification (*e.g.* Leviticus, 14.52; Psalm 51.7). Bitter herbs are mentioned many times; for example, in Exodus, 12.8, where we read that they are to be eaten as part of the Passover meal. The Hebrew word for bitter is מְרֹרִים (mə·rō·rîm) and is not specific to any particular herbs. The key property is bitterness, which symbolises affliction, misery and servitude, and so, in the context of the Passover, represents the suffering of the Israelites under Pharaoh. It can also be seen as a **type** for the sufferings of Jesus.

Sometimes perfume ingredients are mentioned in a context that makes their use in perfume obvious; for example, in the perfume formulae given in Exodus, 30.22–38. In other cases, their use as perfume ingredients has to be implied. One example is labdanum, which is mentioned in Genesis, 37.25 as an item of trade along with other perfume ingredients. Similarly, when the word כְּפָרִים (kə·p̄ā·rîm) is included with a list of other ingredients used in perfume in Song of Songs, 4.13, it is reasonable to conclude that it also is a perfume ingredient. In cases where a word is used only once or twice in the Hebrew and its context implies a perfume ingredient, as is the case with the word כְּפָרִים (kə·p̄ā·rîm) (Song of Songs, 4.13) and the word לֹט (lōṭ) in Genesis, 37.25, it can be very difficult to determine the exact meaning and the usual resort is to consider the Hebrew approach of extending concepts behind a word to enable its use in other ways. The discussions below on labdanum and onycha illustrate this intriguing feature of the Hebrew language, as indeed does the Hebrew word for perfumer/pharmacist (see Chapter 7, Section 7.2 Perfumery or Pharmacy?).

5.3 OTHER FACTORS

There are various factors that lead to uncertainty about exactly what material is intended in each case. Those who translated the Hebrew Bible into the Greek versions, such as the Septuagint, lived at a time when there was a temple in Jerusalem and Jewish priests celebrated the rituals of the Old Testament on a daily basis. The translators might therefore have known what materials were intended, irrespective of linguistic differences. Therefore, the Greek translations might provide important clues. However, the authors and translators of the various books of the Bible were more concerned with theology or history than with perfumery and botany. So, it is possible that they were somewhat relaxed about their choice of the words they used. Also, traders, especially in the natural perfume ingredients business, were not noted for integrity. Falsification of claims and adulteration of materials were all too common up until the advances in analytical chemistry in the latter part of the 20th century AD. Furthermore, those harvesting natural products for extraction of perfume ingredients were not always honest enough, or sufficiently skilled in botany, to ensure that the plant they harvested was what they claimed or thought it to be. One natural products chemist I know, whose role was to analyse the chemical composition of native Australian plants, told me that unless he personally harvested material from the bush, he could not be certain of the exact species, even when presented with intact plant material, such as leaves and flowers, since he needed to see the plant growing in its environment to identify it precisely and accurately. When it comes to attempting to reproduce a Biblical perfume today, we must also bear in mind the fact that plants evolve and so the chemical composition of a 21st century AD sample of a given plant could be different from that of the same species in Biblical times.

5.4 SUMMARY

Bearing all of these things in mind, from language to species identification, the catalogue of Biblical perfume ingredients in Chapter 6 is only my best guess as to exactly what was used. In some cases, the degree of certainty is relatively high but, in others, much less so. I hope that the descriptions make clear which is which. The ingredients are described in alphabetical order of the English names for what I consider to be the most likely substance described by the original Hebrew and/or Greek text and are: agarwood, calamus, cassia, cinnamon, cypress, frankincense, galbanum, labdanum, myrrh, onycha, orris, saffron, spikenard and styrax. Monographs on camphor and henna are also included in the list since they might possibly be intended in place of some other ingredients.

REFERENCE

1. M. Zviely and A. Boix-Camps, *The Israel Chemist and Engineer*, 2015, **1**, 27.

CHAPTER 6

The Ingredients of Biblical Perfumes

6.1 AGARWOOD

The Hebrew word used in Numbers, 24.6, and Proverbs, 7.17, is אֲהָלִים (ă·hā·lîm), and in Psalm 45.8 and Song of Songs, 4.14, it is אֲהָלוֹת (ă·hā·lō·wṯ). It would appear that these are different plural forms of the same word. One English translator (Darby) suggests aloe trees as a translation, but in most English versions, it is given as aloes, aloe wood or lign aloes, all of which terms are still in use and refer to the same substance. It is also known as agarwood and by the Arabic word oud and the Japanese jinkoh. The English word Agarwood is derived from the Sanskrit Aguru. The translators of the Septuagint seem to have had some difficulty with the Hebrew word. In Song of Songs, 4.14, the Septuagint does use the word αλωθ (aloth), but in the other occurrences, it gives quite different translations. The Greek word αλωθ (aloth) seems to be synonymous with αγαλλοχον (agallochon), which appears in the botanical name for the agar-wood tree. In Psalm 44.9 (the Septuagint numbering is equivalent to Psalm 45.8 in other versions), the Greek word is στακτη (stacte), which is Greek for styrax. This is clearly an incorrect translation but at least it does refer to another perfume in-gredient. In Numbers, 24.6, the Septuagint gives σκηναι (skenai),

Perfume in the Bible
By Charles Sell
© Charles Sell 2019
Published by the Royal Society of Chemistry, www.rsc.org

which means tents, and in Proverbs, 7.17, the Septuagint translates it as οἶκόν (oikon), meaning house. In Numbers, the context
clearly implies a tree, and in Proverbs, the word is used in a list of
perfume ingredients so these must have created difficulties in
understanding for readers of the Septuagint and for later translators relying on the Septuagint rather than the original Hebrew.
In the New Testament (John, 19.39), the word used is αλοη (aloe).

The scented substance known as agarwood is the product of
fungal attack of the heartwood of the tree *Aquillaria agallocha*,
which is also known as *A. secundaria* and *A. malaccensis*, and the
history of its use in perfumery goes back several thousand years.
A. agallocha grows in North-East India, particularly Assam, and
parts of Western China. Some related species are also used
and, for example, the product from *A. crassna*, which grows in
Indochina, is known as Eaglewood. The heartwood of *A. agallocha*
is light both in colour and density, but after infection by fungi, it
becomes dark in colour and much denser, and it is this material
that is used in perfumery. Infection by fungi is necessary to
produce the right product for perfumery. Older trees are attacked
by fungi, and trees of any age can be deliberately infected by
drilling a hole and inserting a plug of infected wood. Figure 6.1
shows a tree growing in an agarwood plantation, and the holes
drilled in the bark to enable infection of the wood are clearly
visible. Traditionally, it was considered that three different
families of fungi – *Aspergillus*, *Penicillium* and *Fusarium* – were
implicated and recent research suggests that *Phaeoacremonium
parasitica* is responsible, at least in part. Bacteria and fungi are
often in competition for food, and therefore each tends to
produce chemicals to inhibit the other and eliminate this competition. In pharmacy, some of our best antibacterial agents, *e.g.*
penicillin, were first found in fungi (penicillin was first found in
Penicillium chrysogenum by Sir Alexander Fleming) and, conversely, antifungal agents, *e.g.* nystatin, in bacteria (*Streptomyces
noursei* in the case of nystatin). Thus the presence of fungi in it
helps account for the fact that agarwood has antibacterial properties, although the fungi also induce the tree to produce other
chemicals that may also contribute to the activity. Analysis of
agarwood oil shows it to be a very complex mixture of chemicals.
It has very little in the way of **monoterpenoids** but is rich in
sesquiterpenoids of various classes. The dearth of the more

Figure 6.1 Plantation of agarwood trees (*Aquillaria agallocha*). Reproduced with permission from Givaudan.

volatile **terpenoids** is consistent with its odour being mostly in the end note category. No matter how the active compounds are produced, they make agarwood and the oils it contains useful products for preservation of bodies; in other words, for embalming. This is why on the first Good Friday, Joseph of Arimathea and Nicodemus brought agarwood (aloes) to the tomb to embalm Jesus' body (John, 19.39). A depiction of this can be found in the Corona Chapel of Canterbury Cathedral and is shown in Figure 6.2.

In the Middle Ages, the entombment of Jesus featured in stone memorials as well as glass, and many churches, particularly in France, have an illustration of the scene carved in stone figures, known in France as a Mise au Tombeau. Figure 6.3 shows the celebrated one in the Abbey of Carennac (Lot). Joseph of Arimathea and Nicodemus stand at either end and behind the body of Jesus are Saint John the Evangelist and the women recorded as being present at the crucifixion. One of these is Mary (of Magdala? – see Chapter 9), and in Figure 6.4, in a detail from

Figure 6.2   Jesus' entombment showing Nicodemus and Joseph of Arimathea
applying embalming oils to his body.
Reproduced with permission from the Dean and Chapter of
Canterbury.

Figure 6.3   The Mise au Tombeau at Carennac Abbey.

Figure 6.4 Detail of the Mise au Tombeau at Reygades showing Mary holding a perfume bottle.

the Mise au Tombeau in the church of Reygades (Correze), we see Mary holding a perfume bottle. The Mise au Tombeau in Reygades still shows that the original was painted. The gospels tell us that it was Nicodemus and Joseph of Arimathea who brought perfumes to the entombment, so the bottle that Mary is holding is probably a reminder of the incident at Bethany (see Chapter 9), a device that has become a standard one to identify her in statuary, glass and paintings.

Agarwood can be burnt as an incense or air freshener and the practice nowadays is to use specially designed burners (Figure 6.5). This practice is particularly popular in the Arab world. The odour of agarwood is essentially woody with fruity floral top notes and a sweet balsamic base, reminiscent of sandalwood and with hints of vanilla and musk. Perfumed oils can also be prepared from the wood. An **oleoresin** can be produced by maceration of the wood and soaking it in water. The essential oil can be steam distilled from the wood. The fragrant components can also be extracted using solvents; sandalwood oil, vegetable oil and liquid paraffin have all been used for this

Figure 6.5 Chips of agarwood and an agarwood burner.
Reproduced with permission from Dr Roman Kaiser.

purpose. The product described in John's gospel was probably an extract using vegetable oil or one released by boiling the wood.

6.2 CALAMUS

The Hebrew word קָנֶה (qā·neh) means cane or reed and is used for various types of such plants. One possibility is the bamboo-like *Arundo donax*, which is still used for roofing and for making baskets, flutes and reeds for oboes, bassoons and clarinets. In the book of Job, it is used to describe reeds growing in a marsh (Job, 40.21), Psalm 68 describes cattle roaming in the reeds (Psalm 68.30) and Isaiah tells us that the Servant of the Lord will not break a bruised reed (Isaiah, 42.3). The reference to a marsh in Job would certainly be consistent with calamus. When other words are added as adjectives to קָנֶה (qā·neh), the meaning becomes more specific. So, when the word בֹּשֶׂם (bō·śem), meaning sweet, is added to give קְנֵה בֹשֶׂם (qā·neh bō·śem), then calamus (sweet cane) is intended, as for example, in Exodus, 30.23. The Septuagint translates this as καλαμος ευωδης (kalamos euodēs), as with the Hebrew, literally meaning sweet smelling cane. In the New Testament, when used without the adjective ευωδης (euodēs), meaning sweet smelling, the word καλαμος (kalamos) has various interpretations. When Jesus talks about a reed swaying in the wind, the word for reed is καλαμος (kalamos), and in John's third letter (3 John, 13), the context makes the translation as "pen" the obvious one, presumably indicating a sharpened piece of hollow cane being used as a writing implement; and such tools are still used in calligraphy today. In fact, the Sanskrit word कलम (kaláma) is derived from the Greek and means a pen made from a reed. In Song of Songs, 4.14, the word קָנֶה (qā·neh) appears without qualification but, since it is in a list of perfume ingredients, it can be taken that calamus is the intended meaning. In Ezekiel, 27.19, קָנֶה (qā·neh) is included in a list of three items of trade brought to the port of Tyre by Danite and Greek merchants from Uzal. The other two items are iron and cassia. The latter is a perfume ingredient so, despite the lack of qualifying words, it would again seem that calamus is probably implied. In Jeremiah, 6.20, incense from שְׁבָא (šə·ḇā) Sheba and קָנֶה טוֹב (qā·neh ṭō·wḇ) from a distant land are described as offerings that are unacceptable to the Lord because of the motivation of the givers. Here, the usual English translation is sweet cane but the Septuagint translates it as κιννάμωμον (kinnamomon), which is cinnamon. This is rather odd since the Hebrew word for cinnamon, קִנְּמָן־ (qin·nə·mān), is quite distinct from קָנֶה (qā·neh) and phonetically similar to the Greek. Cinnamon comes from Sri Lanka whereas calamus could be found closer to Israel, and so the comment about a distant land would be more in keeping with the former. Perhaps קָנֶה טוֹב (qā·neh ṭō·wḇ) refers to some particular form of cinnamon stick, but my conclusion would be that its translation as κιννάμωμον (kinnamomon) remains something of a mystery. In Isaiah, 43.24, the word קָנֶה (qā·neh) again appears without qualification and is usually translated as calamus. In this case, God tells the Israelites that they have not used money to buy any קָנֶה (qā·neh) for him and the context of valuable offerings implies calamus. However, the linking to money led the translators of the Septuagint to translate it as ἀργυρίου θυμίαμα (arguriou thumiama), meaning silver incense. Since silver does not burn or have an odour, calamus would seem to be a better translation, and the reference to silver possibly stems from the high price of calamus.

Calamus is the fragrant extract from the rhizomes of the plant *Acorus calamus*. This tall perennial plant bears some resemblance to iris and is known as calamus, sweet cane, fragrant cane, sweet flag, sweet root, sweet myrtle and sweet cinnamon. A flower of calamus is shown in Figure 6.6, and Figure 6.7 shows the base of the stem and the top of the rhizome, which is the part extracted for the oil. It originated in Asia but is now found in swampy areas or growing along brooks, rivers and lakes in the temperate zones of Europe, Asia and America. Currently, the main commercial sources are Eastern Europe, India and Japan. Cultivated plants are propagated by division of the roots. The underground rhizomes of *A. calamus* grow horizontally and are jointed with a spongy texture. They are 2–3 cm in diameter and up to one metre long. Rhizomes of wild plants are harvested in spring or late autumn, washed, freed of root fibres and dried. Although it is usual to dry the rhizomes before extraction since there is not much loss of oil on drying, the fragrant components can also be extracted from fresh rhizomes. Unlike orris, the roots are not peeled before drying since that would result in loss of odorous materials. The fragrant components can be extracted with solvents (vegetable oils in Biblical times) and, nowadays, calamus oil is also extracted by steam distillation.

Figure 6.6   Flower of calamus (*Acorus calamus*).
Reproduced with permission from Dr Roman Kaiser.

Figure 6.7   Base of stem of calamus (*Acorus calamus*).
Reproduced with permission from Dr Roman Kaiser.

The oil has a complex odour, described as a characteristic, warm, woody, spicy odour with great tenacity and a sweet dry out. Since the late 19th century, perfumes with such a profile have been described as "oriental". One pair of closely related chemicals responsible for the distinctive sweet character of calamus is the shikimate derivatives α-asarone and β-asarone. The rich woody elements of the calamus fragrance come from a variety of terpenoids, the intricate molecular structures of which are things of beauty to the organic chemist.

Calamus is one of the components of the sacred anointing oil described in the thirtieth chapter of Exodus. In 15th century France, "Oyselets de Chypre" was a popular incense comprising calamus, labdanum and styrax bound together with tragacanth gum.

Calamus has a long history of use in medicines and flavours. Its use as a medicinal plant is recorded in ancient writings from India and China. The powdered rhizomes were also used as a vermifuge and insecticide in India and the Far East. It is known to have been found in Egypt at least as early as 1300 BC and, although mentioned as part of a poultice for stomach complaints in the Chester Beatty papyrus VI, it was seen primarily as a perfume ingredient in ancient Egypt. It was introduced to Britain in the 16th century and the aerial parts have been used for thatching and as a perfumed floor covering.

It is used in modern herbal medicine as a sedative, laxative, diuretic and carminative, and its antioxidant, antimicrobial and insecticidal properties have been confirmed in modern laboratories. Ingestion can cause hallucinations and its use in foods has been banned because of possible carcinogenic properties. α-Asarone (Figure 6.8) and β-asarone (Figure 6.9) are among the

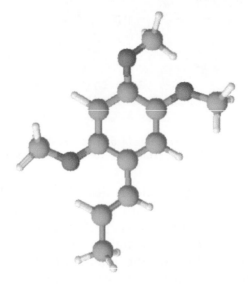

Figure 6.8   α-Asarone.

Figure 6.9   β-Asarone.

chemical components of calamus which might be, at least in part, responsible for some of the biological activity.

Modern research has found that calamus has some **neuroprotective** effects against stroke and the toxic effects of acrylamide.[1,2] It has also been shown to have antioxidant,[3] antimicrobial[4] and insecticidal[4] properties. Its insecticidal properties suggest that it could be useful against cattle tick {*Rhipicephalus (Boophilus) microplus*}.[5] The β-asarone present in calamus has also been shown to reduce lipid accumulation in fat cells.[6]

6.3 CAMPHOR

Camphor is one of the English translations given for the Hebrew word כְּפָרִים (kə·p̄ā·rîm), which appears only once in the Bible, in Song of Songs, 4.13. For example, the King James Version gives camphire, the term then in use to describe camphor.

Camphor is the name given both to the oil extracted from the tree *Cinnamomum camphora* (sometimes referred to as *Laurus camphora*) and to the major chemical component of the oil (Figure 6.10). The tree grows in China, Taiwan and Japan, Taiwan being the most important source of the three. The wood is used for carpentry and building in addition to its use as a source of the oil (Figure 6.11). The oil is distributed throughout all of the

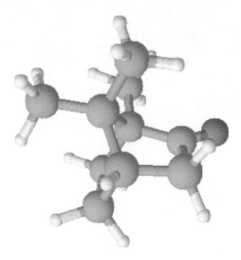

Figure 6.10 Camphor.

Figure 6.11   Camphorwood tree (*Cinnamomum camphora*).
Copyright © Shutterstock.

Figure 6.12   Camphorwood box for linen storage.

wood of the tree. The insect repellent properties of the wood make
it particularly useful for the manufacture of chests for storing
linen; an example is shown in Figure 6.12. The Chinese scientific
writings of Honzo-Komoku in 1596 describe isolation of camphor
by boiling the wood in water and then allowing the camphor
to crystallise from the aqueous extract. So the camphire of the

King James Version (KJV) could have been either an oil or isolated camphor. Nowadays, the oil is extracted by steam distillation.

Whilst it is possible that the word כְּפָרִים (kə·p̄ā·rîm) does refer to camphor, the fact that the Septuagint translates it as κυπρος (kupros) opens up the possibility that the intended species is actually cypress and this will be discussed under that heading.

6.4 CASSIA

There are two Hebrew words that are translated as cassia in English versions of the Old Testament. The first is קְצִיעוֹת (qə·ṣî·'ō·wṯ) and is found in Psalm 45.8. In the Greek of the Septuagint, this appears as κασια (kasia) in Psalm 44.9 (the numbering equivalent to Psalm 45.8 in the Hebrew Bible). It therefore seems clear that this refers to cassia. The other word is קִדָּה (qid·dāh) (Exodus, 30.24), which is translated into the Greek of the Septuagint as ιρις (iris), meaning iris. The interpretation of this word can found below under the heading Orris. The word קִדָּה (qid·dāh) also occurs in Ezekiel, 27.19, in the Hebrew Bible but is omitted in that verse of the Greek of the Septuagint.

Cassia, also known as Chinese cinnamon, refers to fragrant extracts from the evergreen tree *Cinnamomum cassia* (Figure 6.13). The tree is related to that which produces cinnamon and grows in the Kwangsi and Kwangtung regions of South East China, and

Figure 6.13   Cassia tree (*Cinnamomum cassia*).
Copyright © Shutterstock.

also in Vietnam and India. It is also now also cultivated in Japan, Indonesia, Sri Lanka and South America. Like cinnamon, both leaves and bark are used as sources of perfume materials. The dried bark can be powdered and used as spice. Nowadays, steam distillation of leaves from coppiced bush is used to produce the leaf oil. Twigs with a diameter of 4 cm or so are used to produce products from the bark. Since it is the inner bark that is used, all the bark is removed and then the outer layer is scraped off, leaving the inner bark to curl up. Nowadays, a bark oil can be obtained by steam distillation of the bark but this is expensive and production is low. Psalm 45 describes cassia as being used to perfume the king's robes, and so it is likely that it refers to a vegetable oil extract.

Both cassia and the related cinnamon owe their strong, spicy, warm, sweet, woody and balsamic odours to the shikimate chemicals cinnamaldehyde (Figure 6.14) and eugenol (Figure 6.15) that make up a large proportion of the oils. Cinnamaldehyde is responsible for the characteristic odour of cinnamon bark and extracts from it. Eugenol is the major component of clove oil and has an odour strongly reminiscent of clove buds. Cinnamon leaf oil is richer in eugenol than cassia leaf oil, and so the latter has a more cinnamon-like character. Cinnamaldehyde has some anti-bacterial activity, and consequently, so do both cinnamon and cassia extracts.

6.5   CINNAMON

Cinnamon is one of the four components of the sacred anointing oil described in Exodus, 30.23. In this case, there is little doubt as to the identity of the material. The Hebrew word is

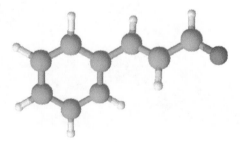

Figure 6.14   Cinnamaldehyde.

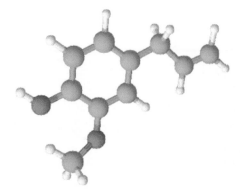

Figure 6.15　Eugenol.

קִנָּמָן (qin·nə·mān) and the Greek translation of the Septuagint is κιννάμωμον (kinnamọmon), both of which words are translated into English as cinnamon, and indeed, the three words all sound similar. The use of cinnamon as a culinary spice is recorded in China in 2500 BC, and it is also mentioned as a herbal medicine in the Ebers Papyrus of 1550 BC (see Chapter 7, Section 7.2 Perfumery or Pharmacy?). Cinnamon was also known to the Greeks and the Romans. It has a history of use in herbal medicine but there is little support from modern research for any medicinal claims.

Cinnamon is extracted from the inner bark of the tree *Cinnamomum zeylanicum*, which grows in Sri Lanka (Figure 6.16). It is a member of the laurel (*Lauraceae*) family, growing to about 10 metres in height and with white blossom. The trees are coppiced, and after three years the shoots are about two to three metres high and about four centimetres in diameter. These are cut and the bark stripped off (Figure 6.17). The outer bark is removed and the inner bark curls up into the sticks that are familiar items in the spice sections of groceries and supermarkets (Figure 6.18). It is best to cut the branches during the monsoon rains when the flow of sap between bark and wood makes separation easier. The story given to Herodotus by Arabian perfumers (see Chapter 7, Section 7.3 Secrecy) suggests that the product he knew was the curled up bark. In Exodus, it is used in the sacred anointing oil, and so this would have been an extract, probably of powdered dried bark, in olive oil. Nowadays, a bark oil is obtained by steam distillation, and a leaf oil is also obtained by steam distillation.

Figure 6.16   Cinnamon tree (*Cinnamomum zeylanicum*).
Copyright © Shutterstock.

Figure 6.17   Stripping cinnamon bark in Sri Lanka.
Reproduced with permission from Dr Robin Clery & Givaudan.

In Biblical times, wild trees would have been used as the source, but in 1770, the Dutch took control of Sri Lanka and replaced wild cinnamon with cultivated plants and controlled trade,

Figure 6.18 Cinnamon bark stripped and dried.

destroying surplus in order to keep prices high. In 1796, England took over Sri Lanka, and the East India Company controlled trade until 1833. Since then, trees have been transplanted to other countries but the oil composition and quality varies due to different soil and climatic conditions. The best and major source is still the Sri Lankan material. The main odour component in the bark is cinnamaldehyde, which comprises 65–75% of the bark oil. The bark oil contains 4–10% eugenol whereas the leaf oil contains 70–80% eugenol. Based on Herodotus' description of cinnamon, it is most likely that the Biblical product would have been extracted from the bark and would have had the characteristic sweet, spicy and tenacious odour of the bark. The use of cinnamon is restricted in modern perfumery because of **skin sensitisation** issues relating to cinnamaldehyde.

6.6 CYPRESS

The Hebrew word כְּפָרִים (kə·p̄ā·rîm) occurs only once in the Bible, in Song of Songs, 4.13. It is translated into English as camphor, henna or cypress and the first two of these are discussed under separate headings. The word used in the Septuagint is κυπροι (kuproi), the plural of κυπρος (kupros) (Cyprus). The modern Greek for cypress is κυπαρισσι (kuparissi), but the use of the plural in the Septuagint and association of cypress with Cyprus suggests that the correct interpretation of the use in Song of Songs, 4.13, is cypress. Various species of trees are mentioned in the Old Testament and cypress is a possible interpretation for one of them. Even if this is correct, it is not impossible that different words were used for the timber and for the oil extracted from the leaves, especially in view of the secrecy surrounding perfume ingredient production.

Cypress oil is the extract from the leaves of *Cupressus sempervirens* (Figure 6.19), a tree that the Greeks and Romans planted on temple and burial grounds, and which is still associated with parks and cemeteries. It grows in Turkey, Iran and Syria, and nowadays, the oil is steam distilled from clippings. However, the components with medicinal properties and those responsible for the amber- and labdanum-like base notes are difficult to extract by distillation whereas solvent extraction gives a richer oil. It has been claimed that this oil, when dropped onto the bedding of affected children, reduces the symptoms of whooping cough. The Biblical oil would probably have been a solvent extract or boiled out of the leaves and skimmed off from

Figure 6.19   Cypress (*Cupressus sempervirens*).
Reproduced with permission from Dr Roman Kaiser.

the water and would have had a refreshing, sweet odour reminiscent of pine needles with balsamic base notes.

6.7 FRANKINCENSE

The Hebrew word לְבֹנָה (lə·ḇō·nāh) is usually translated in the Septuagint as λιβανος (lib'-an-os) and this is the Greek word used in the New Testament also. Both correspond to the English word frankincense, although in modern perfumery, the term olibanum is often used instead of frankincense, especially to describe the oil extracted from the resin. This oil is rich in monoterpenoids and sesquiterpenoids. Some English translations sometimes give incense as the translation for לְבֹנָה (lə·ḇō·nāh). For example, in Song of Songs, 3.6, the Authorised Version and the New English Bible correctly translate לְבֹנָה (lə·ḇō·nāh) as frankincense but the New International Version renders it as incense. I would agree with the Authorised Version and the New English Bible rather than the New International Version in this case.

In Genesis, 2.12, we read that the source of aromatic resins (presumably including frankincense) is the land of חֲוִילָה (ḥă·wî·lāh); Havilah in English. Jeremiah (Jeremiah, 6.20) mentions frankincense specifically and tells us that the source is שְׁבָא (sheb-aw'); Sheba in English. Both Havilah and Sheba refer to the Horn of Africa, Somalia and surrounding countries, including the South West tip of the Arabian peninsula. This region is still the main area of production of frankincense.

Frankincense is the defensive material produced by *Boswellia carterii* (Figure 6.20) and related trees, such as *B. serrata* and *B. sacra* (Figure 6.21), of the *Boswellia* family of the *Burseraceae*. Frankincense collectors start by deliberately cutting the tree and a white emulsion exudes from the wound. This gradually hardens into a pale yellow resin and the collector returns later to harvest the hardened "tears" of frankincense (Figure 6.22). These tears are used directly in incense, and the heat produced when the incense burns vaporises the volatile components of frankincense. Nowadays, an oil is produced by steam distillation of the tears and the dust around them. Various extracts such as **absolutes** and **resinoids** and oils are produced and extraction with ethanol gives a resinoid, which is one of the best fixatives

Figure 6.20   Frankincense tree (*Boswellia carterii*) growing in a greenhouse. Reproduced with permission from Dr Roman Kaiser.

Figure 6.21   Frankincense tree (*Boswellia sacra*) growing in a greenhouse. Reproduced with permission from Dr Roman Kaiser.

used in perfumery. The odour of frankincense is described as having a green top note reminiscent of lemon and apple with a resinous, spicy, woody, balsamic base. It blends well in all perfumes, especially of the "oriental" type.

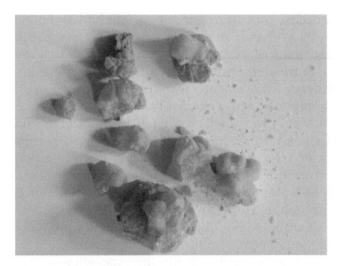

Figure 6.22   Tears of frankincense.

When heated, frankincense gives off a pleasant perfume, and this is how it has been used in incense in religious rituals throughout recorded history. The formula for sacred incense is given in the book of Exodus (Exodus, 30.1–10). Frankincense was to be mixed in equal parts with styrax, galbanum and onycha and the resultant incense burnt on the altar. The incense was also to be used in a censer by Aaron when he entered the holiest part of the tabernacle where the ark of the covenant was kept (Leviticus, 16.12).

Judaism was not alone in use of frankincense for religious purposes and it was used in that way by most ancient peoples. The Israelites imported it through the Phoenicians and also by caravans coming through Persia and Babylon and also across the Red Sea. It is mentioned by many ancient writers, such as Herodotus, Plutarch, Theophrastus, Dioscorides and Pliny. Herodotus tells us that frankincense was the only perfume ingredient of the time that was not used by the Egyptian embalmers when mummifying bodies. The distilled oil of frankincense was known to Valerius Cordus (1515–1544), a German physician and botanist who authored one of the greatest pharmacopoeias and one of the most celebrated herbals in history, and frankincense was included in German apothecary records in 1574 and 1589.

Frankincense was used in many ancient herbal remedies. For example, in Ayurvedic medicine, it was used to treat arthritis, and some modern research supports this property of frankincense,[7] particularly osteoarthritis of the knee.[8] There is also some evidence from modern research that some components of the resin might be active against bladder cancer.[9,10]

The book of Proverbs tells us that "Perfume and incense bring joy to the heart" (Proverbs, 27.9). We know that perfume, like any work of art, can lift our moods, but a specific pharmacological effect of frankincense has now been discovered. Recent research by a team of scientists from the Hebrew University in Jerusalem, Johns Hopkins University and the University of Indiana describes the analysis of the vapours produced when frankincense is burnt. Incensole acetate (Figure 6.23) is among the chemicals they identified and they also showed that it acts on a receptor known as TRPV3. This receptor is found in the skin, where it detects cold and warmth, but it is also found in the brain, where it has a role in anxiety and depression.[11] The research team went on to establish that incensole acetate does indeed reduce anxiety, stress and depression. So, it is likely that the use of incense in religious ceremonies does help to induce a feeling of calm.[12]

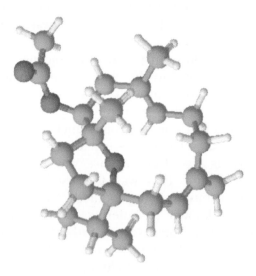

Figure 6.23 Incensole acetate.

6.8 GALBANUM

The Hebrew word חֶלְבְּנָה (ḥel·bə·nāh) and its Greek translation in the Septuagint, χαλβανη (chalbanē), clearly refer to galbanum; for example, in Exodus, 30.34, where it is described as a component of the sacred incense. Galbanum has been known for millennia, both as a perfume ingredient and as a medicinal product; for example, in the Ebers Papyrus of 1550 BC (see Chapter 7, Section 7.2 Perfumery or Pharmacy?). Galbanum has been used as a stimulant, expectorant and antispasmodic. Hippocrates and Pliny both attributed great medicinal properties to it. Dioscorides and Pliny wrote that the original country of origin was Syria. French records report it as being bought from grocers by King John of France when he was a captive in England in 1360 AD, and Venetian merchant records report selling it to London in 1503 AD.

In modern perfumery, galbanum is an exudate from the **umbelliferate** plant *Ferulago galbanifera* (also known as *Ferula galbaniflua*, *F. gummosa* and *F. galbanifera*). This plant, which resembles fennel, cow parsley and hemlock, grows wild in Iran, Lebanon and Palestine (Figure 6.24). Nowadays, commercial

Figure 6.24 Galbanum (*Ferulago galbanifera*).
© Saxifraga – Jasenka Topic. Reproduced with permission, Mr J. van der Straaten.

quantities come from Turkey and Iran. A syrupy gum resin exudes from the aerial parts of the plant and hardens into red-amber tears (Figure 6.25). Steam distillation gives an essential oil, and resinoids are produced by solvent extraction of the tears. Galbanum has a characteristic odour that is aromatic, having a powerful leafy green top note, with earthy, spicy and coniferous notes and a woody balsamic base. The green note of galbanum suggests that it was a component of the "green" incense used in ancient Egypt. In modern perfumery, galbanum is usually considered as a top note because of the green character, but its overall character lasts right through to the end of the dry-out, and so it is also justifiable to consider it a base note in my opinion. This issue simply serves to illustrate the difficulties in trying to classify odour and odorants. The green top note of galbanum was used in "Vent Vert" (Balmain, 1945), and this set in train a fashion for green notes, later examples including "Must" (Cartier), "Chanel No. 19" and "Vol De Nuit" (Guerlain).

It is possible that the galbanum of the Bible was actually obtained from a similar plant, *Galbanum officinale* (also known as *Laserpitium officinale*), which grows across the Middle East from Syria to North Africa. The products would have been similar in use and odour to the modern galbanum.

Figure 6.25   Galbanum resin.

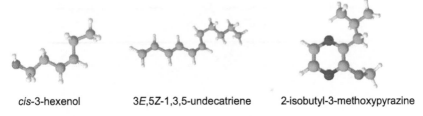

cis-3-hexenol   3E,5Z-1,3,5-undecatriene   2-isobutyl-3-methoxypyrazine

Figure 6.26   Three key green odour components.

In modern perfumery, green notes are usually classified into grass-green, typified by the odour of freshly mown grass (chemically speaking, *cis*-3-hexenol), and galbanum green. Galbanum green itself is due to two main chemicals, 3*E*,5*Z*-1,3,5-undecatriene and 2-isobutyl-3-methoxypyrazine (Figure 6.26). The latter has an odour strongly reminiscent of bell peppers (capsicum), of which it is also a major odour component. The former, 3*E*,5*Z*-1,3,5-undecatriene, has a more grass-like odour and was a puzzle to fragrance chemists for many years because it is an **aliphatic hydrocarbon** and these normally have much weaker and less striking odours. However, it is now known that the odour attributed to 3*E*,5*Z*-1,3,5-undecatriene is actually due to oxidation products of it.

6.9   HENNA

Henna is one of the English translations given for the Hebrew word כְּפָרִים (kə·p̄ā·rîm), which appears only once in the Bible, in Song of Songs, 4.13.

Henna (*Lawsonia inermis*) is a shrub that grows in Sudan, Egypt, Arabia, Iran, India and Sri Lanka (Figure 6.27). A dye is obtained from its leaves and flowers. The flowers are called "flower of paradise" in Arabic. Although the ground dried leaf/flower mixture is primarily used as a dye, it does have a pleasant smell. The leaves have a tea-like herbaceous odour. The flowers have a sweet floral and tea-like odour, the floral notes resembling violet since the oil does contain some chemicals that are also found in violet. However, it is very difficult to produce perfume oil from it and the yield is very low; therefore, it is not commonly used as a fragrance ingredient nowadays. Henna

Figure 6.27   Henna (*Lawsonia inermis*).
Reproduced with permission from Dr Roman Kaiser.

leaves were used in Egyptian cosmetics thousands of years ago, and are still used as a dye for skin and hair, especially in African countries.

In view of all of this, henna seems an unlikely translation for כְּפָרִים (kə·p̄ā·rîm) if a perfume ingredient is intended, as the context would imply. As discussed above under the relevant heading, cypress would seem to be a better interpretation.

6.10   LABDANUM

Labdanum is a physiological exudate of the plant *Cistus ladaniferous*. It has been known throughout history and is mentioned, for example, by Dioscorides, Herodotus and Pliny. Labdanum is sometimes referred to as ledanon or ladanum (not to be confused with laudanum, which is an extract from the opium poppy) and is mentioned in European pharmacopeas of 1589 AD and 1613 AD. Medicinal uses for labdanum included treatment of skin diseases, and in Jordanian traditional medicine, its roots are used as a treatment for bronchitis. Its use in medicines was recorded in ancient Egyptian writings, including as a treatment for dandruff.

C. ladaniferous is a perennial shrub up to two metres high that grows across the Mediterranean region, from Spain to the Greek

Figure 6.28   Cistus (*Cistus ladaniferous*).
Reproduced with permission from Givaudan.

Islands (Figure 6.28). The sticky exudate covers the surface of the leaves and twigs. Boiling of these in water causes the gum to be removed and it can be collected by skimming it off the surface of the water. Nowadays, ethanol extraction of the gum gives a resinoid, steam distillation of which gives an essential oil. Currently, the main production of all of these derivatives is in Spain and all have a sweet balsamic and ambery odour similar to that of the crude gum.

When Herodotus asked the perfumers of Arabia about the sources of their ingredients, he was told a pack of lies. Their story about labdanum was that it was combed from the beards of goats which had been browsing among the cistus bushes. This story persisted into the early 20th century and was particularly prevalent on the islands of Crete and Cyprus. It is possibly true but, to me, the modern practice of boiling in water seems much easier and could well have been the method used in ancient times since boiling was also used for extraction of other oils, as discussed in Chapter 4, Section 4.1.

In ancient times, it was often confused with styrax, myrrh or frankincense and translators of the Old Testament seem to have had the same problem. There are two Hebrew words that have been suggested as meaning labdanum.

The first is שְׁחֵלֶת (šə·ḥê·leṭ), which occurs only once, in Exodus, 30.34. The Septuagint translates this as ονυχα (onycha) onycha, a well known ingredient of incense in ancient times. I think that onycha is the better English translation of שְׁחֵלֶת (šə·ḥê·leṭ) and that it is less likely that labdanum is the intended meaning, my reasons being given in the paragraph below on onycha (Section 6.12).

The second word is לט (lōṭ). This word is found in Genesis, 19, where it is the name of a prominent character in the story of the destruction of Sodom and Gomorrah. Figure 6.29 shows Lot and his family fleeing from the city. The window shows the city buildings collapsing. Lot and his family were told not to look back but his wife did and was turned to a pillar of salt. The medieval glass artist used white glass for the figure of Lot's wife to illustrate this. Otherwise, the word לט (lōṭ) occurs only twice,

Figure 6.29   The destruction of Sodom and Gomorrah.
Reproduced with permission from the Dean and Chapter of Canterbury.

firstly in Genesis, 37.25, and again in Genesis, 43.11. We cannot be certain that לֹט (lōṭ) does relate to a perfume ingredient. Both occasions when it is used occur in the story of Joseph and his brothers. In Genesis, 37.25, the brothers had captured Joseph and put him in a cistern, intending to kill him. The scene is depicted in Figure 6.30, which is of a window in the Corona Chapel of Canterbury Cathedral. However, Reuben pleaded for his life and they decided instead to sell him as a slave to some Ishmaelite merchants. We are given a list of items that the merchants are carrying down to sell in Egypt and one of these is לֹט (lōṭ). The second occurrence of the word is when Joseph's brothers were sent by their father, Jacob, to buy food from Egypt because of a famine in Israel. The meeting is depicted in Figure 6.31, which is a panel in one of Canterbury Cathedral's Bible windows. Jacob suggested that the brothers take some gifts

Figure 6.30 Joseph's brothers put him into a cistern.
Reproduced with permission from the Dean and Chapter of Canterbury.

Figure 6.31   Joseph's brothers in Egypt asking him to sell them food.
Reproduced with permission from the Dean and Chapter of
Canterbury.

to the Egyptian official they will deal with and whom they did not
recognise as being their brother, Joseph. Among the gifts is לֹט
(lōṭ). The context in both cases is consistent with לֹט (lōṭ) being a
perfume ingredient, and most translators take it to be so. The
translation in the Septuagint is στακτη (stactē), that is styrax,
and myrrh is used in the King James Version, the New International Version, the New English Bible and some other English
versions. But the Hebrew for myrrh is מַר (mār) and στακτη is the
Greek for styrax, for which the Hebrew is נָטָף (nā·ṭāp̄). It would
seem that both styrax and myrrh are incorrect translations. Some
English translations, such as that of Darby, give ladanum or
labdanum as translations based on semitic cognates, and I believe that this is more likely to be the correct one. The Hebrew
word לֹט (lōṭ) suggests something that wraps or covers objects,
and so would be entirely consistent with a description of the
gum that covers the cistus plant.

6.11 MYRRH

Myrrh is a term used for extracts from trees of the *Burseraceae* family, and it has 3700 years of recorded use in perfumery and medicine, as for example, in the Ebers Papyrus (see Chapter 7, Section 7.2 Perfumery or Pharmacy?), and myrrh trees from Punt feature in relief drawings of Egyptian tombs dating from a similar era. Many plants which produce odorous chemicals used in perfumery also produce chemicals with a biological activity, and many components of myrrh have both odorous and medicinal properties. Use of myrrh in perfumery and medicine has continued throughout history and across the world, for example, in Ayurvedic and Chinese traditional medicine.

It is therefore no surprise to find references to myrrh in the Bible. The Hebrew word for myrrh is מַר (mār) and the Greek is σμυρνα (smurna), and translations between these two and with the English, myrrh, are consistent. The Bible describes myrrh being used as an anointing oil (Exodus, 30.23), a cosmetic (Esther, 2.12) and an embalming agent (John, 19.39).

Myrrh is an exudate of various trees of the *Burseracea* family. These include *Commiphora abyssinica*, *C. myrrha* (also known as *Balsamodendron myrrha*) (Figure 6.32), *C. Schimperi*, *C. Playfairii*, *C. Hildebrandtii* and *C. serrulata*, the most important being

Figure 6.32 Myrrh tree (*Commiphora myrrha*).
Copyright © Vladimir Melnik/Shutterstock.

C. abyssinica and *C. myrrha*. They grow around the Red Sea coast in Somalia, Sudan and the South West tip of the Arabian peninsula. Their natural habitat is therefore in the same region as the various *Boswellia* species used to produce frankincense but they are also cultivated in parts of Saudi Arabia and Iran.

The oleoresin is produced and held in reservoirs in the trunk. It can exude spontaneously through cracks in the bark but commercial producers stimulate the flow of resin by cutting the bark. The exudate is then bled off and collected and can be air dried. It is usually a viscous yellow to green or brown oil and can harden into reddish brown waxy granular lumps. Nowadays, various **tinctures**, resinoids and absolutes are obtained from the crude material and an oil can be obtained by steam distillation. Distilled oil of myrrh was known to Valerius Cordus (1515–1544), the famous German physician and botanist, and is included in pharmacopoeias and herbals of his time. The odour of all these forms of myrrh is described as being warm, spicy and balsamic.

Many of the chemical components of myrrh possess antibacterial and/or antifungal properties. Among these active compounds are various phenols and members of a group of substances known as sesquiterpenoids, and of particular interest are some belonging to the eudesmane sub-family of sesquiterpenoids. One example of the latter is known as myrrhone (Figure 6.33). These antimicrobial chemicals are produced by the trees to protect them from attack by bacteria and fungi, but these properties are also of use to humans; for example, in embalming

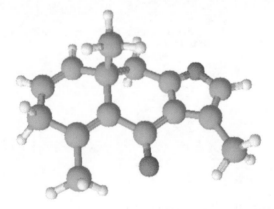

Figure 6.33   Myrrhone.

of bodies. Consequently, it is not surprising that John's gospel tells us that myrrh was one of the substances brought by Joseph of Arimathea and Nicodemus to embalm Jesus' body, as depicted in Figure 6.2 (John, 19.39).

Another member of the eudesmane sesquiterpenoid family is furanoeudesma-1,3-diene (Figure 6.34), and it has been found that it and a related component of myrrh – curzarene (Figure 6.35) – have an action on the opioid receptors in the brain, similar to that of morphine.[13,14] In other words, they serve as pain killers. The combination of antibacterial and

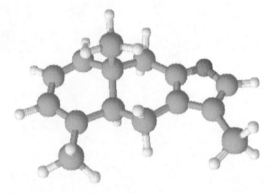

Figure 6.34   Furanoeudesma-1,3-diene.

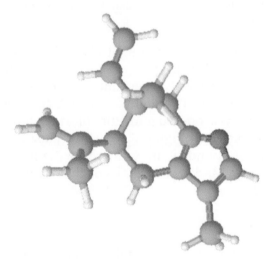

Figure 6.35   Curzerene.

Figure 6.36 Jesus' crucifixion.
    Reproduced with permission from the Dean and Chapter of
    Canterbury.

analgesic properties of myrrh might also explain its use in oral
care products in Ancient Egypt, as recorded in the Ebers Papyrus
(see Chapter 7, Section 7.2 Perfumery or Pharmacy?) and in
dressings for wounds. The ability of myrrh to relieve pain explains why Jesus was offered a mixture of wine and myrrh at His
crucifixion, depicted in Figure 6.36, which shows a panel from
the Redemption Window in Canterbury Cathedral's Corona
Chapel. However, Mark's gospel tells us that He refused to drink
it (Mark, 15.23).

6.12   ONYCHA

In Exodus (Exodus, 30.34) the Hebrew Bible names one of the
ingredients of the sacred incense as שְׁחֵלֶת (šə·ḥê·leṯ). This is the

only occurrence of the word in the Hebrew Bible and its interpretation is perhaps the most contested of any of the Hebrew words used for Biblical perfume ingredients. The Septuagint translates it as ονυχα (onucha), and this meaning is used in most English translations.

The English word onycha is a straightforward translation (in fact a transliteration) of the Greek word ονυχα (onucha) and refers to the operculum of a species of conch or sea snail. The nominative ονυχα should not be confused with the same word, ονυχα, which is the accusative case of ονυξ (onuks) onyx that refers either to the stone, whose name in English is also a direct transliteration of the Greek, or to fingernails or birds' talons. The similarity of the two Greek words has added to the confusion over identifying just what the writers of the original Hebrew intended. The operculum is the covering lid which closes over the animal when it is inside the shell. When removed from the shell, the operculum resembles a fingernail in appearance, and this probably accounts for the use of the Greek word for fingernail to also refer to the incense ingredient. Onycha was burnt as a component of incense in many ancient cultures from the Mediterranean to Eastern Asia. The Baylonian Talmud gives instructions on how to prepare it and improve its quality by treating it with an alkaline plant extract and soaking it in caper juice or white wine. Onycha was so sought after that the snail that was used to make it was fished to extinction. There is a material called dhufran which is used as incense in Oman today. Similarly, a material called nakhla, which in addition to use as incense, is also used in hookah pipes and to flavour tobacco and comes from Yemen, Egypt and India. Both dhufran and nakhla are obtained from sea snails and, therefore, are similar to the onycha of Biblical times.

According to Strong's concordance (Strong's reference 7827; Strong's concordance is a tool for searching and comparing words in the Bible. Nowadays it is easily accessed *via* the Biblehub website, https://biblehub.com), the word שְׁחֵלֶת (šə·ḥê·let) is related to other words suggesting a meaning of either "to roar like a lion" or "to make a sound whilst being peeled off". The latter would certainly fit with onycha. When edible shellfish such as winkles are boiled, there is a popping sound as the pressure inside the shell blows the operculum off the shell to release the steam from inside. Boiling would probably have been used to release the onycha from the molluscs, and a similar popping noise would have been heard. That could well be an explanation for the origin of the word שְׁחֵלֶת (šə·ḥê·leṭ).

Since shellfish are unclean to Jews (Leviticus, 11.9 and 12) as food, some people question whether onycha would be considered suitable for use in the sacred incense. However, the hoopoe was chosen as the state bird of Israel in May 2008 despite being declared unclean in the same chapter of Leviticus (Leviticus, 11.19). Perhaps there has always been a clear distinction between food and general or spiritual cleanliness. Those objecting on the grounds of onycha being unclean suggest labdanum as an alternative translation of שְׁחֵלֶת (šə·ḥê·leṭ). This proposal is based on the fact that שְׁחֵלֶת (šə·ḥê·leṭ) is related to the Hebrew word שָׁחַל (šāḥal), indicating a noise such as the roar of a lion (Hosea, 5.14), and links this to the story of combing labdanum out of goats' beards, the noise then being the vocal protests of the goats. The story of combing labdanum from goats' beards was current at the time of the Arabian perfumers to whom Herodotus spoke, and it is possible that it might have survived through history (it was still being told by producers in Crete and Cyprus in the early 20th century) as a means of protecting the trade secrets of those who actually obtained labdanum by the simpler and more efficient process of boiling cistus plants in water. Since I have an alternative Hebrew word {לֹט (lōṭ)} for labdanum, as described above, I tend to the view that the translators of the Septuagint, who after all would have been practising Jews, knew what they were doing and that contemporary practice gave no objection to ονυχα (onucha) as the correct translation of שְׁחֵלֶת (šə·ḥê·leṭ).

There are a number of other suggestions as to what שְׁחֵלֶת (šə·ḥê·leṭ) referred and my thoughts on them follow, in alphabetical order of the English names.

Amber could refer either to the fossilised brown resin or to grey amber, ambergris, which is an important modern perfume ingredient. In Biblical times, brown amber came from the North Sea coast of Denmark, a source traced by Pythias, a native of the Greek city of Masala which is now known as Marseilles. The source of supply that Pythias traced in the fifth century BC has since been exhausted, and in modern times, amber is more plentiful in the Baltic. However, brown amber has never been used as a perfume ingredient. Ambergris is used in perfumery and, in Biblical times, would have relied on beachcombing for supply. However, ancient documents do not mention it, and it is hard to see a connection between this greyish, waxy substance, washed up on beaches, with either of the derivations of שְׁחֵלֶת (šə·ḥê·leṭ).

Bdellium is the resin derived from *Commiphora mukul*, syn. *C. wightii* a tree of the same family as that from which myrrh is obtained. Bdellium is a resin like myrrh but has quite a different odour, less bitter than that of myrrh. However, it is unlikely to be the substance intended by the word שְׁחֵלֶת (šə·ḥê·leṭ) because bdellium is mentioned in Genesis (Genesis, 2.12), where the word used to described it is הַבְּדֹלַח (hab·bə·ḏō·laḥ). This is translated in many English versions as aromatic resin but the similarity of sound suggests, to me at least, that it could well be the Hebrew for bdellium. It is also difficult to see how bdellium would relate to the likely etymology of שְׁחֵלֶת (šə·ḥê·leṭ). Interestingly, that sentence in Genesis mentions a third substance which is often translated into English as onyx. The Hebrew reads הַשֹּׁהַם (haš·šō·ham) וְאֶבֶן (wə·'e·ḇen), and remember that Hebrew reads from right to left. וְאֶבֶן (wə·'e·ḇen) means stone and הַשֹּׁהַם (haš·šō·ham) is associated with the colour green. If the two words are separate, then הַשֹּׁהַם (haš·šō·ham) might refer to vegetation; if they are connected, then a green stone is implied, hence translations such as onyx or emerald. The Septuagint opts for onyx, as do most English translations. In either case, the Hebrew leaves no doubt that onycha is not intended in this sentence in Genesis.

Benzoin is a resin extracted from *Styrax officinale* and related species such as *S. japonica*. It has been used throughout history and still is; for example, as mentioned in Chapter 8, Section 8.2.1, it is present in Rosa Mystica incense from Alton Abbey. Again, the argument against it being the meaning of שְׁחֵלֶת (šə·ḥê·leṭ) lies in the etymology of שְׁחֵלֶת (šə·ḥê·leṭ).

Clove buds resemble nails, and indeed, in both German and Dutch, the words for clove buds and nails are similar. Since ονυχα (onucha) means nails in Greek, it has been suggested that שְׁחֵלֶת (ū·šə·ḥê·leṭ) refers to cloves. However, since the nails described by ονυχα (onucha) are fingernails and cloves resemble

80 *Chapter 6*

not those but the nails used in carpentry, it is unlikely that cloves is the intended meaning of שְׁחֵלֶת (šə·ḥê·leṯ).

Cuttlefish bone fragments might look something like onycha but they are also derived from shellfish and therefore unclean. Moreover, they are not common ingredients of incense.

Spikenard has been suggested as a translation for שְׁחֵלֶת (šə·ḥê·leṯ) but this is most unlikely since Hebrew consistently uses נִרְדְּ (nir·dî) for spikenard and it is always rendered ναρδος (nardos) in Greek.

Styrax has been suggested as a translation for שְׁחֵלֶת (šə·ḥê·leṯ) but, if this is correct, then one might expect the Septuagint to translate it as στακτη (stakte̲) rather than ονυχα (onucha). Furthermore, both στακτη (stakte̲) and ονυχα (onucha) occur in the same sentence in Ecclesiasticus (Ecclesiasticus = Sirach, 24.15), indicating that the translators of the Septuagint distinguished clearly between the two. As with many of the other suggestions, stacte is difficult to reconcile with the etymology of שְׁחֵלֶת (šə·ḥê·leṯ).

Tragacanth is an odourless and tasteless **polysaccharide** gum obtained from the plant *Astragalus tragacantha* or closely related species such as *A. adscendens*, *A. gummifer* and *A. brachycalyx*. It is used in incense, not as a fragrant component but as a gum to hold other ingredients in a solid form. Its use in incense makes it a possible contender for a translation of שְׁחֵלֶת (šə·ḥê·leṯ) but the argument against this lies in the etymology of שְׁחֵלֶת (šə·ḥê·leṯ).

To summarise, therefore, in view of all the above, I would agree with the most common interpretation; that is that שְׁחֵלֶת (šə·ḥê·leṯ) does refer to onycha.

6.13 ORRIS

The Hebrew word קִדָּה (qid·dāh) occurs in Exodus, 30.24, where it is described as one of the components of the sacred anointing oil. The Septuagint translates it as ιρις (iris) but, despite this, it is often translated into English as cassia, even though both the Hebrew {קְצִיעוֹת (qə·ṣî·'ō·wṯ)} and the Greek {κασια (kasia)} for cassia are quite distinct from the words used in the Hebrew Bible and the Septuagint in Exodus, 30.24. The word קִדָּה (qid·dāh) also occurs in Ezekiel, 27.19, in the Hebrew Bible but is omitted in the Greek of the Septuagint. In Greek, ιρις (iris) can refer either to the iris plant or to a bright circle of light surrounding an object, for instance as in a rainbow. However, when a rainbow is intended by the Hebrew word קַשְׁתִּי (qāš·tî) (as in Genesis, 9.13), the Septuagint translates it as τόξον (toxon), meaning bow as in archery. The use of ιρις (iris), meaning rainbow, is only found twice, both times in the New Testament (Revelation, 4.3 and 10.1). When a bow, as in archery, is intended in the New Testament, the word τόξον (toxon) is used (Revelation, 6.2). Since the Hebrew words are quite distinct, it is doubtful that קְצִיעוֹת (qə·ṣî·'ō·wṯ) and קִדָּה (qid·dāh) would refer to the same plant. In modern perfumery, both bark and leaves of cinnamon are used as raw materials for perfumery and they give different oils. It is therefore possible that, of the two terms translated as cassia, one might refer to a bark extract and one to a leaf extract. However, no mention is made of two different cinnamon products in the Bible, and the leaf oils are nowadays extracted using steam distillation so such a cassia leaf oil could not have been used in Biblical times. Of course, it is possible that a leaf oil was extracted either by boiling the leaves or by using a solvent such as olive oil. In terms of perfumery, either cassia or orris (an extract from iris rhizomes) could make sense. The anointing oil contains cinnamon, which is similar in odour to cassia, but the two are often used to support each other in perfume formulae. However, the use of orris would add a different note and enrich the overall composition. The odorous molecules in orris are closely related in molecular structure to those of violets, and the odour also bears a floral note similar to that of violet flowers. Having discussed this with different perfumers, I am left with the fact that, in modern perfumery thinking, either cassia or orris would make a useful addition to the odour of the anointing oil. In view of my comment above about methods of extraction and mention of only one form of cinnamon, I would therefore consider it more likely that the ingredient described as קִדָּה (qid·dāh) is orris, the extract from iris rhizomes.

Orris is extracted from the rhizomes of the iris species *Iris pallida* and *I. germanica*. Other irises have been used but the best product is that from *I. pallida*, the pallid or Florentine iris (Figure 6.37). Figure 6.37 shows a specimen in Kew Gardens and Figure 6.38 shows one growing at Vézelay Abbey. This iris prefers stony, chalky and lime rich soils on well drained

Figure 6.37   Iris flower (*Iris pallida*).
Reproduced with permission from Dr Rui Fang, Kew Gardens.

slopes, and currently, the main region of production is in Italy, around Florence. The plants are planted in September and harvested three years later, in July or August. Freshly dug rhizomes have a sharp earthy odour but this disappears on storage. The rhizomes are peeled, soaked in hot water then air dried (Figure 6.39). The desired orris odour then develops over two to three years of storage. Various types of product (oils, absolutes, concretes and resinoids) are obtained using solvent extraction and/or steam distillation. Orris concrete is sometimes referred to as orris butter. The absolutes, concretes and resinoids are rich in myristic acid, a fatty acid, and nowadays, this is often removed to give a product that is richer in the volatile odorous components. The most important group of odorant chemicals in orris are the irones, and these are related to the ionones found in violets. Therefore, it is not surprising that the floral and woody odour of orris is reminiscent of that of violets. Figure 6.40 shows molecules of α-irone and α-ionone alongside each other to illustrate the similarity between the two. α-Irone and α-ionone are important odour components of orris and violet, respectively.

Figure 6.38 *Iris pallida* growing at the Abbey of Vézelay in France.

Figure 6.39 Iris rhizomes.

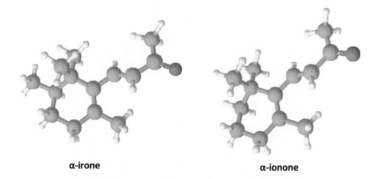

α-irone   α-ionone

Figure 6.40   Molecules of α-irone and α-ionone, showing the similarity of their
structures.

6.14   SAFFRON

Saffron is mentioned once in the Bible, in Song of Songs, 4.14. It
is included in a list of other plants used as perfume ingredients,
and so it is reasonable to conclude that it is also considered to be
a perfume, though it could have been included as a spice or food
colour rather than a perfume. The Hebrew word used is כַּרְכֹּם
(kar·kōm) and the Greek of the Septuagint translates it as κροκος
(krokos), both of which correspond to the English word saffron.

Saffron refers to the spice and perfume extracted from the
stamens of the autumn flowering crocus *Crocus sativus* and also
to the intense yellow dye that can be extracted from them
(Figure 6.41). *C. sativus* is a domesticated species and is probably
derived from the wild plant *C. cartwrightianus*.

C. sativus probably originated in the Bronze age in Crete.
Saffron was known to be traded in Crete 4000 years ago, cultivated in Persia in the 10th century BC and mentioned in Assyrian
writings in the seventh century BC. The Romans brought saffron
to France, and later, the Arabs brought it to Spain. Saffron was
brought to England in the mid-14th century and the town of
Saffron Walden in Essex takes its name from when the town was
a centre of saffron production. English production halted in the
early 19th century but is being reintroduced nowadays. The
largest production at present is in Iran.

Saffron has a long history of use in medicine, as a dye, a food
colouring and a perfume ingredient. Alexander the Great used it
for battle wounds, and Cleopatra used it in her bath. Medicinal

Figure 6.41 The saffron crocus (*Crocus sativus*).
Reproduced with permission from Dr Rui Fang, Kew Gardens.

Figure 6.42 Saffron stamens.

uses included treatment for wounds, scabies, colic, coughs and gastrointestinal problems. The stamens (Figure 6.42) can be used as such in cooking when saffron is used as a colour and flavour for food. Its use in food is particularly characteristic of Mediterranean cuisine. The yellow dye can also be extracted from the stamens and can be used to dye cloth. For example, it is used by Buddhist monks to produce the characteristic yellow colour of their robes. Saffron is rich in **carotenoid** pigments and its intense yellow–red colour is mostly due to the carotenoid pigment crocin. The characteristic odour of saffron, which is described as intensely sweet, spicy and floral-aldehydic, results largely from safranal, and this molecule is a degradation product of the carotenoids found in the stamens. Distillation gives an oil which is unstable and darkens in air. It is mentioned in the taxation records of Nurnberg of 1613 but it is not currently in production. Modern safety concerns mean that saffron is no longer used in perfumery. Comparison of the images of molecules of crocin and saffron in Figures 6.43 and 6.44 will immediately reveal the difference in size of the two molecules. In order to produce a colour, organic molecules (*i.e.* those with structures based on carbon) are usually large and this renders them too poorly volatile to be able to travel through the air and reach the receptors in the nose. Molecules that do reach the receptors and elicit odours are smaller, usually having between 8 and 20 carbons in each molecule and only a limited number of other types of atoms, apart from hydrogen. Thus, crocin is too big to vaporize and reach the nose but it does have an intense colour, and safranal is colourless but elicits a strong odour.

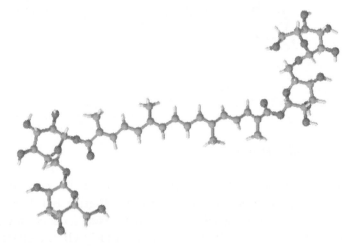

Figure 6.43   Crocin.

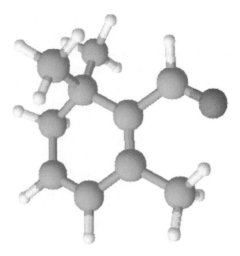

Figure 6.44   Safranal.

6.15   SPIKENARD

Spikenard, also known as nard in English, is mentioned three times in the Old Testament and it occurs in all four gospels. The Hebrew word נִרְדִּי (nir·dî) is used in Song of Songs, 1.12, and means "my spikenard" but is usually translated as "my perfume". In Song of Songs, 4.14, we find נֵרְדְּ (nê·rəd), meaning spikenard, but in the preceding verse, Song of Songs, 4.13, it is נְרָדִים (nə·rā·dîm) which, strangely, is the plural but the Greek translation in the Septuagint is the singular ναρδος (nardos). The gospel writers use both the words μυρος (muros) and ναρδος (nardos). The latter clearly means spikenard, the former is translated either as ointment or perfume. To me, this strongly suggests a concrete of spikenard, which would have the consistency of an ointment and the intense odour of spikenard. In their accounts of what must be the same event, Matthew (Matthew, 26.7) refers to a very expensive perfume, Mark (Mark, 14.3) and John (John, 12.3) to a very expensive perfume that they specify is pure spikenard, Luke (Luke, 7.37) writes simply perfume. After the crucifixion, Jesus' body was treated with myrrh and aloes, as mentioned in the sections on those perfume ingredients, but three of the gospels (Matthew 26.12; Mark, 14.8; John, 12.7) report that, when he was previously anointed with spikenard (see Chapter 9), he said that this was in preparation for his burial. The use of spikenard for perfuming or embalming of bodies was well known and, for example, in Homer's Iliad, it was used on the body of Patroklos. In Hispanic culture, spikenard is used to represent Saint Joseph.

Spikenard is an extract from the roots and young stems of the shrub *Nardostachus jatamansi* that grows at high altitude (3000 to 5000 metres) in the Himalayas. The word jatamansi comes from the Sanskrit meaning hairy since the roots are covered with a profusion of root hairs. The shrub grows to a height of about 60 cm to 1 m tall with basal leaves up to 20 cm long, and its pink, bell-shaped flowers are tiny and clustered. It is a member of the valerian family and is also known as Indian valerian. Like valerian, the roots contain chemicals with powerful action and are therefore used medicinally as well as in perfumery (Figure 6.45). Extracts are used in Ayurvedic medicine and properties attributed to it include tonic, antispasmodic, stimulant, laxative and carminative activity. It was also used as a spice in Ancient Roman and medieval European cuisine. Nowadays, a perfume oil is extracted by steam distillation and it has a heavy, sweet, woody and spicy, animalic odour reminiscent of valerian, ginger, cardamom and Atlas cedarwood. It is still an expensive ingredient, as in Biblical times.

Figure 6.45   Roots of spikenard (*Nardostachus jatamansi*).

6.16 STYRAX

In terms of language, the identity of styrax is straightforward; the Hebrew is נָטָף (nā·ṭāp̄) and the Greek of the Septuagint translates it as στακτη (stakte̞), and both words translate into English as styrax. The Septuagint tends to translate other Hebrew words also as στακτη (stakte̞) but these can be identified as mistranslations. However, there are three trees that are used nowadays to produce a product known as styrax. These products are all similar in terms of method of production, odour and perfumery use. The most likely of the three to be the source of Biblical styrax is *Liquidambar orientalis*. Another possibility is *Styrax officinale* but this tree grows in Eastern Asia and Indochina whereas *L. orientalis* is native to Turkey. The third is American styrax, *L. Styraciflua*, which is used in modern perfumery but clearly would not have been available to the people of the Bible. In all, there are about 130 species of trees in the *Styracaceae* family whose members can be found in the warm temperate zones right across the Northern hemisphere and also in South America. A number of them are used nowadays to produce styrax and a resin known as benzoin (Figure 6.46). To the chemist, this might be a little confusing since benzoin is the name of a key chemical component of the balsam. Benzoin resin is used medicinally, for example in Friar's balsam.

Since *L. orientalis* is the most likely source of the styrax described in the Bible, it is the one described below. Figure 6.47 shows an example of this tree growing in Hadlow College in Kent, and for comparison, a specimen of *Styrax japonica*, a relative of *S. officinale*, growing in a garden in Cornwall is shown in Figure 6.48.

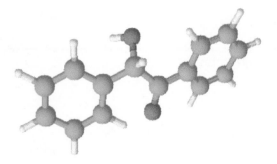

Figure 6.46 Benzoin.

Figure 6.47   Styrax (*Liquidambar orientalis*) growing at Hadlow College, Kent.

Figure 6.48   Styrax (*Styrax japonica*) growing in a garden in Cornwall.

The cover design of the book shows another variety, *Styrax obassia*, in flower.

Styrax is mentioned as a component of incense in Exodus, 30.34. Some English translations call it storax or gum resin, others simply transliterate the Greek (stacte) and yet others mistranslate it as myrrh. The use of styrax in incense has continued throughout history, the famous "Oyselets de Chypre" of 15th century France being one example. Styrax is a pathological exudate from the sapwood of *L. orientalis*. It was once known as Levant styrax but this is a misnomer since it did not come from the Levant but instead from South West Turkey. The tree is a member of the *Hamamelidaceae* and grows to heights of up to 20 metres. Harvesting of styrax begins when the trees are about three or four years old. Patches of bark and sapwood are cut from either side of the tree in May and the exudate is scraped off every few days until mid-November. This process is repeated annually until the diameter of the tree is reduced to only a few centimetres, after which the tree is left for three or four years to recover. Eventually, the tree is felled for firewood. The exudate is collected in pots and boiled in water to release the product known as styrax. This is an opaque semi-liquid that is white to brownish in colour and smells strongly balsamic, resembling the scent of Peru balsam and benzoin. Steam distillation gives a pale yellow to dark brown oil. Styrax is rich in cinnamic acid and esters of it and also contains cinnamyl alcohol, phenylpropanol and 1-phenylethanol. The last of these is also known as styrallyl alcohol, and it dehydrates on heating or on exposure to acid to give styrene (Figure 6.49),

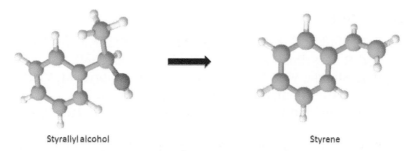

Styrallyl alcohol   Styrene

Figure 6.49 Conversion of styrallyl alcohol to styrene.

much better known as a **monomer** for the production of polystyrene. So, styrene – a chemical that is generally thought of as a typical modern, man-made product – was actually there as a natural product in perfumes of Biblical times.

The Hebrew name for styrax, נָטָף (nā·ṭāp̄), is particularly interesting, as is the Greek word στακτη (stakte). In Hebrew, נָטָף (nā·ṭāp̄) means something that is caused to drip and the Greek στακτη (stakte), when used as an adjective (here in the feminine form), also means dripping and is derived from the verb σταζω (stadso), meaning to drip. Therefore, in both languages, the name for styrax describes exactly how the exudate is removed from the tree. Cutting into the sapwood causes the resin to drip out into the pots of those harvesting it. In addition to using the word to describe styrax, it is also used in the sense of dripping or dropping: "the earth trembled and the heavens dropped" (Judges, 5.4), "my words dropped upon their ears" (Job, 29.22), "the heavens dropped at the presence of God" (Psalm 68.8), "for the lips of a strange woman drop as a honeycomb" (Proverbs, 5.3), "Your lips drop as the honeycomb" (Song of Songs, 4.11), "my hands dropped with myrrh" (Song of Songs, 5.5), "lilies dropping sweet smelling myrrh" (Song of Songs, 5.13), "drop your word toward the south" (Ezekiel, 20.46), "drop your word toward the holy places" (Ezekiel, 20.46), "the mountains shall drop down new wine" (Joel, 3.18), "do not drop your word against the house of Isaac" (Amos, 7.16), "the mountains shall drop sweet wine" (Amos, 9.13). The two uses of נָטָף (nā·ṭāp̄) in Ezekiel and the first one in Amos link the dropping of words to prophesy, and Micah uses נָטָף (nā·ṭāp̄) to mean prophesy (Micah, 2.6 and 2.11).

REFERENCES

1. P. K. Shukla, V. K. Khanna, M. M. Ali, R. Maurya, M. Y. Khan and R. C. Srimal, *Hum. Exp. Toxicol.*, 2006, **25**(4), 187.
2. P. K. Shukla, V. K. Khanna, M. M. Ali, R. Maurya, M. Y. Khan and R. C. Srimal, *Phytother. Res.*, 2002, **16**(3), 256.
3. D. S. Asha and G. Deepak, *J. Herbs, Spices Med. Plants*, 2011, **17**(1), 1.
4. R. K. Balakumbahan, K. Rajamani and K. Kumanan, *J. Med. Plants Res.*, 2010, **4**(25), 2740.

5. S. Ghosh, A. K. Sharma, S. Kumar, S. S. Tiwari, S. Rastogi, S. Srivastava, M. Singh, R. Kumar, S. Paul, D. D. Ray and A. K. Rawat, *Parasitol. Res.*, 2011, **108**(2), 361.

6. M.-H. Lee, Y.-Y. Chen, J.-W. Tsai, S. C. Wang, T. Watanabe and Y.-C. Tsai, *Food Chem.*, 2011, **126**(1), 1.

7. C. Drahl, *Chem. Eng. News*, 2008, **86**(51), 38.

8. K. Sengupta, K. V. Alluri, A. R. Satish, S. Mishra, T. Golakoti, K. V. S. Sarma, D. Dey and S. P. Raychaudhuri, *Arthritis Res. Ther.*, 2008, **10**, R85.

9. M. B. Frank, Q. Yang, J. Osban, J. T. Azzarello, M. R. Saban, R. Saban, R. A. Ashley, J. C. Welter, K. M. Fung and H. K. Lin, *BMC Complementary Altern. Med.*, 2009, **9**, 6.

10. M. G. Dozmorov, Q. Yang, W. Wu, J. Wren, M. M. Suhail, C. L. Woolley, D. G. Young, K.-R. Fung and H.-K. Lin, *Chin. Med.*, 2014, **9**, 18.

11. A. Moussaieff, N. Rimmerman, T. Bregman, A. Straiker, C. C. Felder, S. Shoham, Y. Kashman, S. M. Huang, H. Lee, E. Shohami, K. Mackie, M. J. Caterina, J. M. Walker, E. Fride and R. Mechoulam, *FASEB J.*, 2008, **22**(8), 3024.

12. A. Moussaieff, M. Gross, E. Nesher, T. Tikhonov, G. Yadid and A. Pinhasov, *J. Psychopharmacol.*, 2012, **26**(12), 1584.

13. Analysis of myrrh reviewed, L. O. Hanus, T. Řezanka, V. M. Dembitsky and A. Moussaieff, *Biomed. Pap.*, 2005, **149**(1), 3–28.

14. P. Dolara, G. Moneti, G. Pieraccini and N. Romanelli, *Phytother. Res.*, 1996, **10**, S81.

CHAPTER 7

Perfumery

7.1 STRUCTURE OF MODERN PERFUMES

In modern perfumery, the individual ingredients or notes are usually classified as top, middle or bottom notes; or, more romantically, as head, heart and base notes (Figure 7.1). The distinction is essentially based on volatility, the top notes being the most volatile, and therefore the first to escape from the liquid into the air and thus the first to be perceived. The middle notes constitute the heart of the perfume, its basic core, and the base notes are those that persist longest after application to a surface such as skin, hair, paper or furniture. This is why choice of a perfume to purchase should not be made on the first impression on opening a bottle but after allowing it to "dry out" on a perfumer's blotter or on the skin. That first impression will represent mostly the top note and a full appreciation of the perfume can only be gained after smelling it over time on dry out. Perfumers of the 21st century have access to both natural and man-made ingredients, and the latter are more easily classified because they are usually much purer and often consist of a single chemical entity. On the other hand, natural perfume ingredients are very complex mixtures of different chemicals produced by the plant. Most essential oils consist of hundreds of different plant chemicals. The odour profile of the oil therefore changes during dry out and the intensity at each stage depends

Perfume in the Bible
By Charles Sell
© Charles Sell 2019
Published by the Royal Society of Chemistry, www.rsc.org

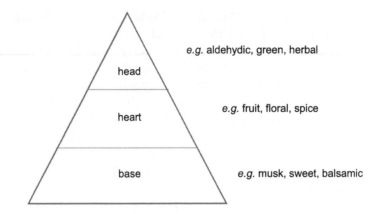

Figure 7.1 Structure of a modern perfume.

not only on the volatility of individual chemicals but also their intrinsic intensity of odour. Galbanum is an example of confusion caused by this. Galbanum is usually classified as a top note because its intense "green" odour is easily distinguished in the top note of any perfume containing it. However, that green note persists long into the dry out and is still discernible days after application onto a perfumer's blotter. When perfumers are being trained, they will smell perfumes and essential oils, especially those with a more interesting dry-out such as ylang ylang and lavender, over a period of hours or days in order to see the different components in them.

When a modern perfumer, perfume evaluator or critic describes a perfume, it will be done in terms of the notes from which it is constructed. An appropriately named modern example to quote in the context of Biblical perfumes is Angel. Both of the following descriptions of Angel indicate a well-balanced blend of top, middle and end notes. The proportions of the three are usually something like 25% top note, 50% middle note and 25% end note.

The manufacturer of Angel describes it in the following way. Angel has a sparkling top note of bergamot and dewberry building onto an aqueous floral note enhanced by rose. The base is a powerful blend of patchouli enveloped by honey, chocolate, caramel and vanilla.

The perfume critic, Michael Edwards, in his book *Perfume Legends*, picks out much the same notes, although his classification into top, middle and end notes differs. He describes Angel as having a celestial top note of bergamot and jasmine, a delicious heart of red berries, dewberry and honey and a voluptuous soul of patchouli, vanilla, chocolate and caramel.

A visit to a department store or specialist perfume shop will quickly reveal that many modern perfumes have aldehydic and/or green top notes and a very sweet base. In the middle of the 19th century, organic chemists began to be able to make substances, some identical to natural chemicals and others that did not exist before, in the laboratory. This art is known as organic synthesis and skills in it developed throughout the 20th century and on to the present day. The availability of pure, relatively inexpensive ingredients from laboratories and factories transformed the perfume industry. The new generation of ingredients gave the perfumer a more secure source of supply, lower cost, superior performance in use and, what is a surprise to the layman, in many cases, safer products to replace natural ingredients that were found to have undesirable properties such as **skin sensitisation.** These new ingredients led to one of the big differences between Biblical and modern perfumes – cost. Nowadays, everyone who wishes to can buy and use perfume to the extent that only the wealthy were able to in Biblical times.

Probably the best known modern fragrance is Chanel No. 5. This perfume was created by Ernest Beaux and launched in 1921 by Gabrielle (Coco) Chanel, who initially gave bottles of it to customers in her couturier shop. She then decided to sell it in the shop, and this started the fashion of couturiers also selling perfume; a fashion that continues today. The other fashion started by Chanel No. 5 was a family of fragrances known as al-dehydic fragrances. Ernest Beaux used synthetic **aldehydes** (first made in 1877) to give a novel top note to his creation, adding a new interest to its heart of rose and jasmine oils. The aldehydes he used have waxy, fresh air odours with citrus aspects. A typical modern aldehyde of this type is 2-methylundecanal, shown in Figure 7.2. These ingredients were not available in Biblical times.

In Chapter 6, Section 6.8, the grass-green top note chemical *cis*-3-hexenol was mentioned. This chemical is responsible for the scent of freshly mown grass but it could not be extracted from grass since it only forms during the cutting process and is too volatile and unstable to isolate using ancient methods.

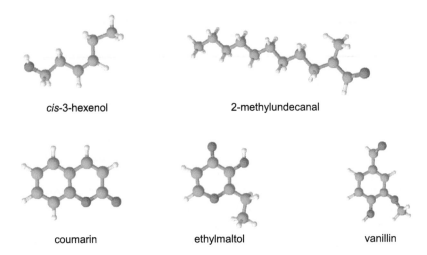

cis-3-hexenol   2-methylundecanal

coumarin   ethylmaltol   vanillin

Figure 7.2   Some modern perfume ingredients unavailable in Biblical times.

Thanks to modern chemistry, we can now make a **nature iden-tical** version, and this is used to give that characteristic top note to modern fragrances, something else that was impossible in Biblical times. A molecular model of *cis*-3-hexenol is shown in Figure 7.2.

Figure 7.2 shows three of the ingredients responsible for the characteristic sweet base note of modern perfumes: coumarin, ethylmaltol and vanillin. Coumarin is responsible for the sweet aroma of new mown hay and the richest natural source of it is from Tonka beans. Extraction from hay is not practicable, and Tonka (*Dipteryx odorata*) is a South American species and therefore unknown to perfumers of Biblical times in the Western hemisphere. Tonka beans are expensive, and so it was a boon to perfumers when nature identical coumarin was first synthesised in 1868 by the chemists Perkin and Graebe. The perfumer Paul Parquet, also part owner of the perfume house Houbigant, used it in his perfume Fougère Royale, launched in 1882. Fougère is French for fern but the perfume does not contain anything re-lated to ferns. Nonetheless, it started a perfume fashion, known as fougère, just as Chanel No. 5 created the family of aldehydic fragrances. Vanillin is the key odour component of vanilla (*Vanilla planifolia*), another South American native and thus unknown to perfumers of Biblical times. It does occur in many other plant species but only the vanilla bean serves as a useful commercial source. Nature identical vanillin appeared in 1877 and was used, together with coumarin and heliotropin, in Guerlain's perfume Jicky, another important trend setter. Heliotropin is the major odour component of the Peruvian species *Heliotropium arborescens* and closely related plants. So, it is another perfume ingredient that would have been unknown to Western and Middle Eastern perfumers of Biblical times, except as a minor component of plant extracts. Nature identical heliotropin was first made in 1869.

Figure 7.2 shows five perfume ingredients that are characteristic of many modern perfumes but would only have been present in trace quantities in substances accessible to Biblical perfumers. When I searched the Bible for ingredients, it was no surprise at all to find only natural ingredients since it was only towards the end of the 19th century that organic chemistry developed to the stage of being able to design and manufacture new ingredients for perfumery. What was surprising was that the perfume ingredients that are mentioned in the Bible are all long lasting base notes. Therefore, Biblical perfumes would be quite different from modern perfumes with their heady top notes. Biblical perfumes would have derived their character predominantly from base and would have been long lasting. One could speculate on the possibility of a symbolism of something fundamental or everlasting, but perhaps this is a step too far. The perfume ingredients described by Herodotus are also all base notes. An alternative explanation might lie in the techniques available. The extraction methods of Biblical times are more suited to base note materials, and storage and transport of the more volatile top and middle note ingredients would also have presented difficulty.

7.2 PERFUMERY OR PHARMACY?

Modern fragrance chemists do their best to develop chemicals with an odour but no adverse effects to humans, animals and the environment. In other words, they look for substances with only one biological activity – stimulation of olfactory receptors. As already mentioned in Chapter 3, Section 3.2, the odorous chemicals that plants and micro-organisms produce are usually designed to perform some other role, such as defence against bacteria, fungi or herbivorous insects and animals. That role is one that requires them to have some biological activity. This can result in safety issues or can endow the chemicals with properties that are useful in herbal medicines.

As mentioned in the Introduction (Chapter 1, Section 1.1), the Hebrew word רֹקֵחַ (rō·w·qê·aḥ) is somewhat ambivalent and can indicate either perfumer or pharmacist, as evidenced by different translations into English. In fact, in Biblical times and right up to the present, the professions of fragrance chemist and pharmaceutical chemist are closely related. The scientific techniques that I used in the research laboratory whilst working at Givaudan are the same as those used at the nearby pharmaceutical giant, Pfizer. The Hebrew word רֹקֵחַ (rō·w·qê·aḥ) comes from a stem meaning to mix or stir. Those making perfumes in Biblical times and their contemporaries who made herbal medicines both would have taken plant materials and mixed and stirred them together to produce the desired product, either perfume or medicine. In modern research, organic chemists design molecules that they hope will have the desired effect on receptors in the body. The perfume chemist aims at the olfactory receptors in the nose whilst avoiding effects on other receptors, whereas the medicinal chemists will target those other receptors, such as hormone receptors. In fact, the olfactory (smell) receptors in the nose and hormone receptors in other organs belong to the same class of receptors – a group known as 7-transmembrane G-protein coupled receptors (GCPRs for short). For example, as mentioned earlier, some of the odorant molecules in myrrh act on olfactory receptors in the nose and also on the opioid receptors in the brain. Some olfactory receptors are found elsewhere in the body and perform different roles there. A number of odorant receptors have been found in parts of the digestive tract and are thought possibly to be involved in appetite control or as signals for release of digestive proteins. One olfactory receptor is produced in high levels in prostate cancer cells, though why this should be is, as yet, unknown. The pharmaceutical industry can afford to spend much more on research than the fragrance industry can, and so much of what we know about odour receptors comes from extensive studies on their relatives, such as the adrenalin and opioid receptors, of interest to the pharmaceutical industry.

Many of the chemical components of natural perfume ingredients possess antibacterial and/or antifungal properties. Therefore, such ingredients have been used throughout history as antimicrobial agents. For example, the Ebers Papyrus, an Egyptian herbal dating from about 1550 BC, records the use of a recipe containing myrrh, cinnamon and galbanum for freshening the breath. Not only do these substances have pleasant smells, they all contain antibacterial chemicals which will kill the bacteria responsible for bad breath. The antibacterial properties also make them useful as embalming agents. Herodotus reported that the Egyptians used myrrh, cassia and every other fragrant substance, except frankincense, when mummifying corpses. As described above, modern research has confirmed not only the antibacterial property of myrrh but also its analgesic (pain killing) property, and frankincense has been shown to reduce anxiety and stress. The Ayurveda, the Hindu text, also contains many accounts of herbal medicines.

The profession known in Hebrew as רֹקֵחַ (rō·w·qê·aḥ) is mentioned in three places in the Bible. In Exodus, 30.35, Moses gives instructions to " ....make a fragrant blend of incense, the work of a perfumer", and in 1 Samuel, 8.13, the prophet warns the Israelites that, if they insist on having a king, he "will take your daughters to be perfumers and cooks and bakers". In both of these cases, the context implies perfumer rather than pharmacist and, in both cases, the Septuagint gives μυρεψικος (murepsikos = perfumer) as the translation. However, in Nehemiah, 3.8, either meaning of רֹקֵחַ (rō·w·qê·aḥ) would be possible since it refers to Hananiah, who is the son of a member of the profession. The Septuagint dodges the decision about translation by simply rendering ρωκεῖμ (rō·w·qê·îm) as a phonetic representation of the Hebrew plural. Interestingly, many translations into English and other modern languages ignore the word for son in both the Hebrew Bible and the Septuagint and write that Hananiah himself was a perfumer/ pharmacist. In the Septuagint, in the book of Ecclesiasticus (also known as the Wisdom of Sirach) (Ecclesiasticus, 38.8), the context of μυρεψικος (murepsikos) leaves no doubt that the intended meaning is pharmacist since it talks about his making medicines that the doctor can use.

In this chapter and in Chapter 6, there are numerous references to perfume ingredients being used as herbal medicines, and this is not at all surprising, in view of the discussion in Chapter 3, Section 3.2. Recently, there has been a move to use odours and perfumes in other ways than for the specific physiological activity of one or more of their chemical components. Dogs have been used throughout history for tracking and for finding truffles. In recent years, there has been a growing trend to train dogs to detect, by smell, certain cancers and the onset of hypoglycaemic attacks and epileptic seizures. This approach relies on the dogs' ability to recognise and interpret small changes in a complex mixture of odorants; in the case of cancers, these would mostly be unpleasant odours. However, pleasant odours and perfumes are also being used in diagnosis of various human conditions. Differences in detection versus description of odours might prove useful in planning treatment for those suffering from Alzheimer's disease, and odour tests are helpful in diagnosis of the various forms of Parkinson's disease. Research has shown that memories can be triggered by words, pictures or smells. Memories triggered in elderly people by words or pictures tend to be memories from their teenage years, whereas those triggered by smell are more likely to have been formed in the first 10 years of life. Research has therefore looked into smell as therapy, including studies into induction of sleep, relaxation and stress reduction during medical procedures. This is all consistent with the assertion in Proverbs that perfume induces happiness (Proverbs, 27.9). More details about the use of perfume in these ways can be found in Chapter 4 of *Chemistry and the Sense of Smell*.

Chapter 3, Section 3.2 also explains why, contrary to popular opinion, modern safety issues have at least as great an effect on natural perfume ingredients as on synthetics (man-made ingredients). For example, the fifth amendment to the European Cosmetics Directive notes 26 perfume ingredients that dermatologists suspect might be skin allergens. Of these, 19 (approximately three quarters) are natural chemicals present in essential oils and other extracts. Many natural materials, including many of those listed above as Biblical perfumes, are restricted in use in modern perfumery. For instance, the presence of cinnamaldehyde and eugenol in cassia and cinnamon limits the amount of these oils that can be used in perfume.

7.3   SECRECY

Until the development of modern analytical chemistry in the middle of the 20th century, secrecy was a major feature of the perfume industry. Perfumers, and those who supplied them with raw materials, used secrecy as the predominant way of protecting their trade secrets, hence their occupations and hence their livelihoods. Nowadays fragrance chemists can take a perfume and subject it to an analytical technique known as gas chromatography/mass spectrometry; GC–MS for short. This technique separates out all the individual chemical entities in the mixture and produces a molecular "fingerprint" of each. These "fingerprints" can then be compared against a library of those of known chemical compounds and so each can be identified. Thus, in the space of an hour or so, the detailed chemical composition of a perfume can be determined. However, that detail is just a list of all the different chemicals present and their relative proportions. It takes an experienced analytical fragrance chemist to work out what the original formulation might have been. For instance, the **monoterpene** limonene is a component of a vast number of natural oils and extracts, although it is particularly important in citrus oils. So, a reasonable level of limonene might indicate the presence of a citrus oil in the formula, and the chemist will then examine the GC–MS results to identify if that oil was from orange, mandarin, lemon, grapefruit or bergamot. Modern perfumers and fragrance chemists rely on patents to protect their work. In Biblical times, the safest way to protect the trade secrets of perfumery was to ensure that no-one else knew the sources of the raw materials used, the methods of extracting them from natural sources, the techniques involved in blending them into perfumes and, of course, the formula of the perfume.

When compiling his *Histories* in the fifth century BC, Herodotus (484–425 BC), the Greek historian and geographer, travelled to Arabia, which was then a centre of the world perfume industry, and asked the perfumers there about their sources of raw materials. To protect their trade secrets, they told him a pack of lies. They told him that cassia was a plant that grew in a shallow lake and was protected by ferocious winged creatures and that those harvesting it needed to protect their eyes from the animals. They said that no-one knew where cinnamon grew but that the sticks were used by large carnivorous birds to build their nests. The nests were constructed on high ledges on steep cliffs that no-one could climb. The collectors of cinnamon placed large joints of meat at the foot of the cliffs and the birds were so greedy that they always carried too much meat up to their nests, which then broke under the weight and fell to the ground where the sticks could be picked up. As described above, cassia and cinnamon sticks are actually the curled up bark of trees that grow on firm ground in China and Sri Lanka, respectively. The Arabians told Herodotus that labdanum was combed out of goats' beards. This is just about possible but, also as described above, simply boiling the plant material in water is a much easier process. The story of combing labdanum from goats' beards persisted into the 20th century AD in Crete and, I suspect, was no more truthful then than in the fifth century BC. The Arabians also told Herodotus that labdanum was the main component of incense. That is likely to be another lie since that role was, and still is, usually played by frankincense.

CHAPTER 8

Perfume in the Bible

8.1 THE PERFUMES OF MOSES

The pervasive secrecy of the perfume industry means that published perfume formulae are rare. It is therefore a little surprising to find two in the book of Exodus. In that case, protection of the formulae was provided not by secrecy but by threat of punishment. Reading the various punishments prescribed in the Torah, the Jewish law books, it would seem that the most severe punishment after the death penalty was that the offender should be "cut off from his people". This punishment was meted out to those guilty of crimes such as incest (Leviticus, 20.17) and sacrificing children to the Canaanite god Molech (Leviticus, 20.5). It was also the punishment for anyone who made up either of the perfumes for any use other than in the tabernacle (Exodus, 30.33 and 30.38).

After leading the Children of Israel out of Egypt (depicted in Figure 8.1, a panel from one of Canterbury Cathedral's Bible Windows), Moses climbed Mount Sinai to meet with God and there was given the Ten Commandments. He was also given The Law and detailed instructions on how to build a tabernacle in the desert and how to conduct worship there. The two perfume formulae were among these instructions (Exodus, 30.22–38).

Perfume in the Bible
By Charles Sell
© Charles Sell 2019
Published by the Royal Society of Chemistry, www.rsc.org

Figure 8.1 Moses leading the Children of Israel out of Egypt.
Reproduced with permission from the Dean and Chapter of Canterbury.

Taking the tabernacle instructions as a whole shows that worship there must have been a wonderful sensory experience engaging all five senses.

The sacred anointing oil contained 500 shekels of myrrh, 250 shekels of cinnamon, 250 shekels of calamus and 500 shekels of what was either orris or cassia, all dissolved in a hin of olive oil as the solvent (Exodus, 30.23–24). In modern measures, this gives a formula as follows. However, remember the serious punishment given for making this perfume for personal use; neither the author nor the publisher can accept any responsibility for anyone doing so.

Sacred anointing oil

5.75 kg of myrrh
2.875 kg of cinnamon

2.875 kg of calamus
5.75 kg of orris or cassia?
4 litres of olive oil

This oil was used by Moses to anoint the Ark of the Covenant and the tabernacle with all its contents and utensils (Exodus, 30.26). Moses also used it to anoint Aaron and his sons as the first Jewish priests (Exodus, 30.30). This is depicted in a panel of the Redemption Window in the Corona Chapel of Canterbury Cathedral and shown here as Figure 8.2. The quantities given indicate a large scale operation of perfume blending resulting in over 20 kg of anointing oil.

Figure 8.2 Moses using the holy oil to anoint Aaron as the first Jewish priest. Reproduced with permission from the Dean and Chapter of Canterbury.

The sacred incense contained equal amounts of styrax, onycha, galbanum and frankincense, ground together to a fine powder (Exodus, 30.34). This was the incense to be used in front of the Ark of the Covenant (Exodus, 30.36), as depicted in a panel of the Redemption Window in the Corona Chapel of Canterbury Cathedral and shown here as Figure 8.3. Later instructions included one that, on the Day of Atonement (Yom Kippur), Aaron should take a censer with coals from the altar, add two handfuls of the incense and burn it in front of the Ark of the Covenant (Leviticus, 16.12–13). When Nadab and Abihu, Aaron's sons, willfully carried out these instructions incorrectly, they died (Leviticus, 10.1–2). Similarly, only descendants of Aaron were permitted to use the sacred incense as directed by Moses and 250 people who broke this decree were consumed by fire (Numbers, 16.35 and 16.40). So, again, as for the sacred

Figure 8.3 Aaron entering beyond the veil of the tabernacle with a censer containing the holy incense. Note the Ark of the Covenant in front of Aaron.
Reproduced with permission from the Dean and Chapter of Canterbury.

anointing oil, remember the serious punishment given for making this incense for personal use; neither the author nor the publisher can accept any responsibility for anyone doing so. However, we also read that Solomon burnt incense thrice annually in order to fulfil temple obligations (1 Kings, 9.25). Perhaps this was a different incense or perhaps a priest descended from Aaron burnt it on Solomon's behalf. Certainly, one of King Uzziah's wrong actions leading to contracting leprosy was to burn incense in the temple (2 Chronicles 26.16 and 26.19). Uzziah was King of Judah from about 785 BC to about 740 BC, the leprosy episode being dated to about 750 BC. The books of Kings and Chronicles record the wrong deeds of a number of the Israelite kings, and there are many instances where incorrect use of incense is listed among the wrong actions. Conversely, good kings such as Hezekiah, who ruled Judea from about 715 BC to about 686 BC, took action against those who used incense wrongfully (2 Chronicles, 30.14).

Hezekiah's great-grandson, Josiah, was born about 648 BC and reigned as King of Judah from about 640 BC to about 609 BC. Among Josiah's many reforms, the books of 2 Kings and 2 Chronicles record that he broke down altars where incense had been offered idolatrously (2 Kings, 23.5 and 23.8 and 2 Chronicles, 34.4 and 34.7). During the rebuilding of Jerusalem after return from exile in Babylon, Tobiah, an Ammonite, had been allowed to use a room previously used for storing, among other things, the sacred incense. Nehemiah, who had been put in charge of the rebuilding by King Artaxerxes of Persia, knew that Mosaic law dictated that no Ammonite should be allowed into the holy sites of Israel (Deuteronomy, 23.3) and had Tobiah removed, the room purified and the incense replaced in it (Nehemiah, 13.1–9). In the last book of the Hebrew Bible, God asserts that incense will be brought to him by people of all nations (Malachi 1.11).

8.2   USES OF PERFUME IN THE BIBLE

8.2.1   Sacred

Incense formed an important part of Jewish ritual in the Old Testament. Incense is a compounded perfume, not to be confused with frankincense, which is one ingredient often used in incense. The Hebrew for frankincense is לְבֹנָה (lə·bō·nāh), which corresponds to the Greek λιβανος (lib'-an-os), although sometimes the Septuagint gives the incorrect translation of στακτη (stacte̱), which is the Greek for styrax. When incense is the intended meaning, Hebrew uses the words קְטֶר (qiṭ·ṭēr) or קְטֹרֶת (qᵉṭō·rĕṭ). Sometimes one piece of Hebrew text will use one of these when another uses the other. As mentioned in Chapter 5, sometimes the word is qualified, as for example, when the word מִקְטָר (miq·ṭār) is added to indicate burning on an altar or in a censer. We must judge from the context if the incense referred to is the holy incense of Exodus, 30.34, or some other formulation. In some cases, the context might make this clear, and sometimes a variant of the word is used, as for example, in Ezekiel, 23.41, where קְטָרֶת (qə·ṭā·rə·t) is used to indicate that the incense is God's incense, presumably that of Exodus, 30.34.

The books of Exodus and Leviticus give instructions about the use of incense in rituals, initially in the tabernacle in the desert but these practices continued in the temple after its construction in Jerusalem. Incense was burnt both on the altar and in censers, and Leviticus tells us that the aroma from the incense burning on the altar is pleasing to the Lord (Leviticus, 2.2). Moses' instructions were that the altar should be made of acacia wood overlaid with pure gold (Exodus, 30.1–3) and placed in front of the Ark of the Covenant in the Tabernacle (Exodus, 40.5). This is depicted in Figure 8.3. Aaron was to burn incense on it every morning and evening when he tended the lamps, an ordinance that was to continue for the generations to come (Exodus, 30.7–8). No other incense, grain offering or burnt offering was to be burnt on that altar (Exodus, 30.9). Leviticus gives an instruction that incense – in this case, frankincense is specified as the incense – should also be added to grain offerings to give a pleasant odour to the burnt grain (Leviticus, 2.1). The use of censers as a means of heating incense to vaporise it is mentioned in Leviticus (Leviticus, 16.12) and this is possibly copied from Egyptian practice. The instruction for Aaron to burn incense in a censer in front of the Ark of the Covenant on the Day of Atonement {וֹם כַּפֻּר (yō·wm ḵap·pêr) Yom Kippur} was to produce a cloud that would obscure the Mercy Seat on top of the Ark and therefore prevent Aaron from seeing God and dying (Leviticus, 16.12–13).

This scene is depicted in one of the panels in the Redemption window of the Corona chapel in Canterbury Cathedral where it serves as a type of the Ascension (Figure 8.3). In Isaiah's vision (Isaiah, 6.1–5), despite the cloud of incense, he saw God and feared death because of his guilt but an angel touched his lips with a burning coal from the altar to purify him so that he would live to carry out his commission (Isaiah, 6.6). In Aaron's time, it was he who tended the altar and used incense, but later, the priests instituted a rota system for this duty (1 Chronicles, 24.1–19). Zechariah was fulfilling his duty in that rota when an angel appeared to him to tell him that he and his wife Elizabeth would be the parents of John the Baptist (Luke, 1.8–11). The daily rituals continued after Jesus' death and resurrection (Hebrews, 10.11) until the destruction of Herod's temple by the Romans in 70 AD.

When plague struck the Israelites in the desert, Aaron took a censer of burning incense as an atonement for them and by his standing between the living and the dead with the censer of incense, this brought the plague to an end (Numbers, 16.47).

The instructions given in Exodus for worship in the tabernacle concern much more than the use of incense. Chapters 25 to 30 of Exodus give instructions for making the tabernacle and the ornaments, vestments and everything in it. There was copious use of gold, silver and bronze with different gemstones and fabrics of wool, goat hair and linen in colours such as red, scarlet, blue and purple. There were incense and scented oils and, no doubt, music. In Chapter 10 of the book of Numbers, there are instructions for the use of silver trumpets in worship. These trumpets would probably have been of similar design to those found in Egyptian tombs of the same period. The worship in the tabernacle would have been a genuinely multi-sensory experience. The perfumes would have pleased the sense of smell; the different colours of wool and other fabrics, the sense of sight; the music, the sense of hearing; the different textures of the fittings and fabrics, the sense of touch; and the sacrificial food, the sense of taste.

When the Israelites settled in Jerusalem, Solomon built a magnificent temple there. The details of the building of this temple are given in 1 Kings Chapters 5–8 and 2 Chronicles Chapters 2–4. It used stone and various scented woods, certainly cedar and possibly sandalwood, for its construction and, like the tabernacle in the Sinai desert, it contained stimuli for all of the senses. In addition to perfumes for the sense of smell, the colours and shapes would have pleased the eyes, music the ears and the many different textures the sense of touch. Worship in the tabernacle and the temple would have been truly multi-sensory experiences. Modern research shows how interaction between the senses is very important, congruent signals from different senses producing a much more intense effect than those of the individual senses. As explained in Chapter 2, the sensory images formed in our minds that we call smell, sound, taste, touch and sight are all constructed in the brain and each is affected by the other, as demonstrated by the inability of experts to correctly describe the taste of a white wine to which a tasteless red dye has been added. So, worship in the tabernacle and temple would have been an all-embracing experience with a profound effect on the worshippers.

Through Isaiah (Isaiah, 43.24), God complained to the Israelites that they were not approaching Him in worship with calamus to offer as incense but, when it was offered in meaningless rituals without a proper motivation, the message was very different. In Leviticus, God had warned the Israelites that idolatrous practices and burning incense to idols would bring down a severe punishment on them (Leviticus 26.30–31). So it is not surprising that both Isaiah (Isaiah, 1.13) and Jeremiah (Jeremiah, 6.20) told the Israelites that hypocrisy is abhorred and empty ritual descried. There is no point in religious meetings and practices, including burning incense to God, if their hearts are not in the right place and in the right relationship with God. Jeremiah equated burning incense to other gods with murder, theft, adultery and perjury (Jeremiah, 7.9) and also blamed the people's burning of incense to other gods for the destruction of Jerusalem by the Babylonians (Jeremiah, 11.17 and 32.29 and 44.23). Hosea, Ezekiel and Jeremiah all pointed out repeatedly that idolatrous use of incense contributed to the fall of both Israel and Judah and the destruction of Jerusalem with subsequent captivity in exile in Babylon. Hosea compared Israel's behaviour to that of his unfaithful wife (Hosea, 2.13) and pointed out that the more God called to them, the more they persisted in their idolatrous use of incense (Hosea, 11.2). Habakkuk complained to God that he seemed to be leaving unpunished successful, but idolatrous, fishermen who burned incense to their drag nets (Habakkuk, 1.16). Ezekiel compared Israel's unfaithfulness to that of a prostitute when they burnt the sacred incense and the food God had provided for them to eat, to idols (Ezekiel, 16.18–19). He told them that they would lie slain in the places where they had done this (Ezekiel, 6.4 and 6.6 and 6.13) and again he reminded them that it was God's incense that they had offered to the Baals (Ezekiel, 23.41–48). Similarly, in the 44th chapter of his book, Jeremiah pointed out how burning incense to the Queen of Heaven was one of the idolatrous practices that led to the fall of Jerusalem (Jeremiah, 44.15–28). However, through Ezekiel, God compares the reformed Israelites returning from exile to fragrant incense (Ezekiel 20.41).

The prophets of the Old Testament are shown in the north **oculus** of Canterbury Cathedral. The design of the entire window represents the Old Dispensation, the Law and the Prophets, the core of Judaism. The central roundel can be seen in Figure 8.4.

Figure 8.4 The central roundel of the north oculus of Canterbury Cathedral. Reproduced with permission from the Dean and Chapter of Canterbury.

The central square shows Moses, representing the Law, and the Synagogue, representing the Jewish people. The four major prophets surround them, reading clockwise from top right are Jeremiah, Daniel, Ezekiel and Isaiah. In the corners between them are the four cardinal virtues; reading clockwise from the top, justice, temperance, fortitude and prudence. The window dates from about 1178 but was heavily restored by Samuel Caldwell Junior in the early 20th century. The contrasting image in the south oculus is described in Chapter 9.

Incense does not appear to have played a significant role in the early Christian church. It is mentioned in the book of the Revelation of Saint John, where it is taken to symbolise the prayers of the saints, as for example, in Revelation, 5.8 and 8.3–4. Just as incense rises upwards, so the prayers of the saints rise up to heaven. This reflects back to Psalm 141.2, where prayers are also likened to incense. In Revelation, 5.8, the context mentions 24 elders, implying both the Jewish (12 patriarchs) and Christian (12 apostles) peoples, and so possibly indicates incense use in Christian worship. The apostle Paul spoke of Jesus' sacrifice as a fragrant offering (Ephesians, 5.2) and he described gifts from his friends in a similar way (Philippians 4.18). However, major use of incense in Christian ritual developed in the fourth century, probably copied from secular practice of the time. It was certainly established by the Middle Ages, and medieval windows do depict images such as the censing angel, shown in Figure 8.5, which is one of a pair now in the Great South Window of Canterbury Cathedral to which they were moved from a window in the nave. Of course, such images might refer to Biblical scenes, such as that of Revelation, 8.3, rather than to contemporary practice.

Incense is still used today in Roman Catholic, Greek Orthodox, Russian Orthodox and in some Anglican churches. One incense used nowadays, for instance in Canterbury Cathedral, is called "Rosa Mystica" and is made at Alton Abbey in Hampshire. The formula is secret, of course, but the material safety information data sheet (a legal requirement for all perfumes and perfume products) indicates that it contains frankincense and other resins (such as benzoin and tolu balsam) in addition to essential oils. As might be expected from its name, the label also suggests the presence of rose oil because of the inclusion of geraniol in the list of chemicals present in it. Geraniol is a major component

Figure 8.5   A censing angel.
Reproduced with permission from the Dean and Chapter of Canterbury.

of rose oils and extracts. Based on what we find in the Old Testament, the floral notes added by the rose oil would not have been found in the incense used in the temple in Jerusalem.

The use of perfumed oil for anointing in religious rituals has continued from Old Testament practice, as described above for the perfumes of Moses, into Christian practices. In Orthodox, Roman Catholic, Anglican and Lutheran churches, perfumed anointing oil is used at baptism and/or confirmation as a seal of the covenant of baptism, and it is also used in admission to holy orders. It is also used to anoint English monarchs as part of the Coronation ritual and, being considered the holiest part of that service, is conducted out of sight of all but the monarch and the

Archbishop of Canterbury. French kings were traditionally anointed in the same way in the cathedral of Reims. The oil used for Christian baptism and anointing of monarchs is known as chrism. This word is derived from the Greek χρισμα (chrisma), meaning the act of anointing. The Greek title "Christ" is derived from the same root, and both it and the Hebrew title "Messiah" mean "The Annointed One".

There is currently some interest in the smells associated with spaces, such as buildings, and the effects of these smells on memory and emotion. So, is there a "smell of churches"? In my experience, each church has its own environmental odour, but two main odour components stand out. The first I associate with old churches, where heating is a luxury and there is a distinct smell of damp, particularly in crypts. The key odour chemical responsible for this is geosmin (Figure 8.6). Geosmin is also responsible for the smell of rain falling on earth that has been dry for some time and it is the scent that animals follow to find water, since geosmin is a fungal product and only occurs where there is water. Camels are so sensitive to it that they can detect water sources from many kilometres distance. The other odour that I associate with churches is that of frankincense. If incense is used in a church, frankincense seems to provide the longest lasting notes. Smell is known to produce effects, often unconsciously, on the memory and the emotions. Therefore, the smell of a damp church or a church where incense has been used is likely to conjure up memories from the past or trigger emotional responses. These will be dependent on the life experience

Figure 8.6 Geosmin.

of the person smelling the odour. In *Learning to Smell*, Wilson and Stevenson show how smell is learnt and so individual experience will determine whether or not a specific memory or emotion will be triggered. Similarly, in *The Great Pheromone Myth*, Doty makes a convincing case that in mammals, reaction to smell is a conditioned response and even those responses thought to be pheromones, akin to those of the insect world, are not innate but rather are conditioned responses more akin to the salivation response of Pavlov's dogs to the sound of a bell. In other words, someone who finds that the smell of a church conjures up a specific memory, or induces emotions such as unease or calm, will be experiencing this because of a, probably unconscious, association with something in their past.

One environmental odour that was certainly present in ancient Egypt (mentioned in the Chester-Beatty papyrus) and persisted into medieval Europe was that of using calamus as a floor covering. The calamus reeds would have provided a softish dry covering for the floor that could absorb dust or mud from footwear, and the pressure exerted by feet walking on it would have released the sweet odour of the plant. It is therefore quite likely that this practice would also have been found in Jewish and early Christian temples, churches, synagogues and such places of worship of the time.

8.2.2 Secular

The writers of the Old Testament were well aware of the value of perfume and its role in everyday life. They noted how perfume brings joy to the heart (Proverbs, 27.9), and when Isaiah warned the Israelites of the hardships they would endure in exile in Babylon, along with sackcloth instead of fine clothes, he said that perfume would be replaced by a stench (Isaiah, 3.24).

The use of perfume for embalming bodies is well known as far as ancient Egypt is concerned, but the practice was clearly also employed in Israel. We read that the body of Asa, King of Judah who died about 873 BC, was embalmed using perfume (2 Chronicles, 16.14).

As discussed in an earlier chapter, when oil is mentioned, it might be either perfumed or not and, unless there is a qualifying adjective or prefix, we must guess from the context which it is.

Isaiah (Isaiah, 61.3) describes the oil of happiness replacing mourning so, with Proverbs, 27.9, in mind, we might conclude that this is perfumed oil. The oil of gladness is also mentioned in Psalm 45.7. The psalmist (Psalm 133.2) uses precious oil as a metaphor for the peace when brothers live together harmoniously and also relates it to eternal life. In what is perhaps the most famous psalm of all (Psalm 23.5), anointing the head with oil is one of the many blessings listed by the psalmist. In these two psalms, it would seem likely that the oils in question are perfumed, just as was the oil that Moses used to anoint Aaron (Exodus, 30.30).

In modern life, we use perfume in many different ways, not only as fine fragrance for application to the skin but also in cosmetic creams and lotions, soap, laundry powder, air fresheners and many other everyday commodities. Some of these commodities, such as shower gels and kitchen and bathroom cleaners, did not exist in Biblical times, but in those that did, we read that perfumes were used just as they are today. Nowadays, we use laundry detergents and fabric conditioners as a way of perfuming clothes. Those products might not have been known in Biblical times but clothes and bedding were perfumed. We read in the Song of Songs that the beloved's clothes have a perfume like that of Lebanon (Song of Songs, 4.11) and the psalmist tells us that the king's robes are perfumed with myrrh, aloes and cassia (Psalm 45.8). We learn in the book of Proverbs that bedclothes were perfumed with such ingredients as myrrh, aloes and cinnamon (Proverbs, 7. 17). The punishment for using the holy incense for personal use (Exodus, 30.38) implies that incense was used as an air freshener in houses and palaces. This is confirmed by Ezekiel when he describes the idolatrous behaviour of the Israelites, who sat on elegant couches before tables on which they placed the holy oil and burnt the holy incense (Ezekiel, 23.41–48). The practice of using calamus as a floor covering was known from ancient times and would have been another form of room air freshener.

8.2.3 Personal Perfume

Chapter 2 summarises the way in which human body odour is produced. Most of us prefer to remove and/or cover up our personal odour and many use perfume for this purpose. This practice is certainly not new and there are many references to it in the Bible. Bearing in mind that many perfumes were extracted from plants using olive oil or animal fat, it is fairly safe to assume that mentions of oils and lotions would mean perfume or perfumed cosmetic preparations. As an example of how God cares for his people, the Psalmist tells us that He "anoints our heads with oil" (Psalm 23.5). During a three week period of mourning and serious contemplation, Daniel tells us how he refrained from using lotions, the implication being that he normally did use them (Daniel, 10.2–3). Figure 8.7 shows Daniel in the lions' den in Babylon. It was also in Babylon that Daniel interpreted a dream for King Nebuchadnezzar, with the result that Nebuchadnezzar honoured Daniel by, among other things, presenting him with incense (Daniel, 2.46). Figure 8.8 depicts Amos, who lived about 150 years before Daniel and described

Figure 8.7   Daniel in the lions' den.
Reproduced with permission from the Dean and Chapter of Canterbury.

Figure 8.8 Amos.
Reproduced with permission from the Dean and Chapter of Canterbury.

fine cosmetic preparations as one of the signs of luxurious living of the people of his time and warned them that they should pay less attention to those things and more to their religious heritage (Amos, 6.6). In the Song of Songs, we learn that the king wears perfume (Song of Songs, 1.3) and that it is made from myrrh and all the spices of the merchant (Song of Songs, 3.6). His beloved also wears perfume and she says that her hands were flowing with myrrh when she opened the door for him (Song of Songs, 5.5).

8.2.4 Courtship

One of the most beautiful of all love poems is the Song of Songs. It describes the courtship of a king and his beloved. Its inclusion in the Bible is justified because it is seen as an allegory of God's relationship with His people. The King in the poem is generally considered to be Solomon. We know that perfume was important in his life because of the many references, as described in various places above, to the perfumed woods used in his temple and the perfumes used on his body and clothes. When the Queen of Sheba came to visit him, she brought many gifts. In one verse (1 Kings, 10.10) that lists these gifts, there are two Hebrew words, וּבְשָׂמִים (ū·bə·śā·mîm) and כַּבֹּשֶׂם (kab·bō·śem), which are both usually translated into English as spice and the Greek of the Septuagint gives ηδουσμον (hedousmon), meaning mint, for both. The structure of the Hebrew words implies something sweet, so they might refer to spice or perfume. Since Sheba was a source of perfume ingredients, I would like to think that perfumes were among the gifts that the Queen brought to Solomon. In Canterbury Cathedral, the window panel shown in Figure 8.9

Figure 8.9 The Queen of Sheba presents gifts to Solomon.
Reproduced with permission from the Dean and Chapter of Canterbury.

is placed in a typological window next to a panel depicting the magi giving their gifts to the infant Jesus (Chapter 1, Figure 1.1). Thus, we know that the medieval artists saw the Queen of Sheba's gifts to Solomon as a type of the presentation of gifts to Jesus by the magi.

His beloved tells us that Solomon's perfume was a reason that all the maidens loved him (Song of Songs, 1.3), whilst he says that her perfume is more pleasing than any spice (Song of Songs, 4.10). She describes him as being like a sachet of myrrh resting between her breasts (Song of Songs, 1.13), perhaps indicating one way in which perfume was worn.

The role of perfume in courtship was also important in the Persian Empire. When the Persian king Xerxes was looking for a bride, he was presented with a selection of young women from which to choose. Before being taken to meet Xerxes, each girl underwent 12 months of beauty treatment involving a variety

Figure 8.10 Esther from the Powell window in Bath Abbey (1947).

of different perfumes and cosmetics (Esther, 2.12). He chose
Esther, an Israelite woman famed for her beauty. This proved to
be a fortunate choice for all of the Israelite exiles since it was
Esther's brave intervention that saved them from the persecution
initiated and carried out by the wicked Haman, one of Xerxes'
officials in the Persian court. The story is told in the book of
Esther, and Figure 8.10 is a picture of Esther from a window in
Bath Abbey.

Perhaps the most significant account of courtship in the Bible
starts with a woman called Naomi, as recounted in the book of
Ruth. The latter is portrayed in a window in Bath Abbey
(Figure 8.11). Naomi and her husband Elimelech came from
Bethlehem. When famine struck their country, they went with
their two sons, Mahlon and Kilion, to live in the land of Moab.
Elimelech died there, and Mahlon and Kilion married Moabite
girls, Orpah and Ruth. Later, both Mahlon and Kilion died and

Figure 8.11   Ruth from the Powell window in Bath Abbey (1947).

Naomi decided to return to Bethlehem. She urged her widowed daughters-in-law to remain in Moab and to remarry to Moabite men. Orpah did so but Ruth refused to leave Naomi, declaring that Naomi's people would be her people and Naomi's God, her God. So, Naomi relented and took Ruth back to Bethlehem with her. Naomi decided to do some match making and identified a relative of hers as a potential suitor for Ruth. His name was Boaz. Naomi gave Ruth advice on how to court Boaz, and this advice included wearing her best clothes and her best perfume (Ruth, 3.3). It worked. Boaz fell in love with Ruth and married her. They had a son called Obed and he, in turn, had a son called Jesse. Jesse's youngest son, David, was chosen by God and anointed King of Israel by the prophet Samuel. The individual ancestor portraits shown in Figures 8.12 to 8.15, are from

Figure 8.12   Boaz.
Reproduced with permission from the Dean and Chapter of Canterbury.

Figure 8.13 Obed.
Reproduced with permission from the Dean and Chapter of Canterbury.

Canterbury's unique Ancestors series dating between 1174 and 1207.

In Jewish law, marriage with a Moabite was strictly forbidden, and no descendant of a Moabite, down to the tenth generation, was allowed to enter the assembly of the Lord (Deuteronomy, 23.3). Despite this, David, who was a third generation descendant of the Moabite Ruth, was chosen to be the greatest of Israel's kings and the prophesied ancestor of the Messiah. In John's revelation (Revelation, 22.16), Jesus says he is both the root and the branch of Jesse's tree. The importance of the prophecies telling of the Messiah's descent from Jesse (*e.g.* Isaiah, 11.1) and David (*e.g.* Isaiah, 9.7) was not overlooked by medieval artists. Therefore, many churches and cathedrals have "Jesse Tree" windows showing Jesus' ancestry from Jesse. In these windows,

Figure 8.14 Jesse.
Reproduced with permission from the Dean and Chapter of Canterbury.

Jesse is usually shown in a recumbent position at the bottom of the window and a tree is shown sprouting from his abdomen or groin area. Panels above Jesse then depict selected descendants on the branches, in chronological order, with Jesus at the top. Figure 8.16 shows the bottom panel from such a Jesse tree in Canterbury Cathedral. All but two of the original panels have been lost, so the panel in the figure is a copy based on the style of the two surviving panels and probably those of medieval French cathedral windows. Figure 8.17 shows the more elaborate Jesse tree in the church of Saint Mary Magdalene in Troyes.

One can speculate on how great a part was played by Ruth's bottle of perfume in leading to Israel's greatest king (David), its wisest (Solomon), a whole dynasty of kings of Judah and, eventually, to Jesus.

Figure 8.15   David.
Reproduced with permission from the Dean and Chapter of Canterbury.

8.2.5   Seduction

The apocrypha is not recognised as a canonical text by either Judaism or protestant Christianity. It does, however, contain some interesting historical material, particularly in the books of the Maccabees. The book of Judith is so littered with elementary errors of history and geography that it is difficult to accept it as historical fact. However, the basic story is an entertaining one. Judith was a widow. Her husband Manasses had died of sunstroke (Judith, 8.3). The devout Judith had worn sackcloth for three years and four months after her husband's death when she heard of the danger to Israel from the Assyrian army led by Holophernes. She decided to take matters into her own hands and threw off her widow's weeds, dressing in the finest clothes

Figure 8.16 Jesse from the Jesse tree in Canterbury's Corona Chapel. Reproduced with permission from the Dean and Chapter of Canterbury.

she had and applying perfume to her face. Thus attired, she set out to seduce Holophernes. Her ruse worked and Holophernes invited her to a banquet. He didn't notice what Judith had concealed in her skirts, and during the meal, she drew out her husband's sword and cut off Holophernes' head. The ensuing panic in the Assyrian troops allowed the Israelite army to drive them back and free Israel from the danger of invasion.

8.2.6 Perfume – The Tool of the Prostitute's Trade

Looking again at the quotation about perfuming the bed (Proverbs, 7.10–20), but now putting it into its context, we see that it is from a metaphorical passage. It is an allegory for the struggle between wisdom and folly and describes how a young man is seduced by a prostitute. The prostitute represents folly. The woman entices the young man by telling him that her

Figure 8.17 The Jesse tree in the church of Saint Mary Magdalene, Troyes.

husband is away, and she describes how she has adorned her bed with the finest coloured linen and used perfumes to add to its appeal. This clearly makes perfume out to be a tool of the prostitute's trade.

CHAPTER 9

Perfume at Bethany

9.1 THE GOSPELS

The four gospels lie at the heart of the New Testament since they record the life, death and resurrection of Jesus. In medieval Christian art, their writers are often represented in a **tetramorph**. The south oculus window in Canterbury Cathedral depicts the Christian church. In the central roundel, shown in Figure 9.1, Jesus is at the left of the central square with the first and last letters of the Greek alphabet, A (alpha) and Ω (omega), behind him showing that he is the first and last, the beginning and end of all things. On the right, a woman depicts the Christian Church with the Latin text ECCLESIA (a direct translation of the Greek), meaning church, behind her. Hence, the central square alludes back to the Song of Songs where bridegroom and bride represent God and mankind, as described in Chapter 1. This theme of marriage between God and his people runs across the Bible, for example in Hosea's marriage (Hosea, Chapters 1–3) and in the parable of the wise and foolish virgins (Matthew, 25.1–13), right through to John's Revelation at the end of the Bible where the bridegroom/bride metaphor corresponds to Jesus and the Christian church. The layout of the whole window reflects that of the north oculus (see Chapter 8, Section 8.2.1), showing the change from the Old Dispensation (the Law and the Prophets) to the New. Around the central square of the south oculus is the

Perfume in the Bible
By Charles Sell
© Charles Sell 2019
Published by the Royal Society of Chemistry, www.rsc.org

Figure 9.1   The central roundel of the south oculus of Canterbury Cathedral. Reproduced with permission from the Dean and Chapter of Canterbury.

tetramorph of the four **evangelists**, the writers of the four gospels. Reading clockwise from the top right, an eagle represents Saint John, a winged lion Saint Mark, a winged bull Saint Luke and a winged man Saint Matthew. In the corners between these are the Christian virtues of (reading clockwise from the top) hope, love, humility and faith. This window, like that of the north oculus, dates from about 1178, but in this case, the restoration was carried out by George Austin Junior in 1850.

9.2   COMPARISON OF ACCOUNTS

Apart from the accounts of Jesus' baptism and subsequent temptation, the feeding of the five thousand and Jesus' passion and resurrection, there is only one event recorded in all four gospels (Matthew, 26.6–13; Mark, 14.3–10; Luke, 7.36–50; John, 12.1–11). This concerns a dinner at which Jesus is the guest of honour. During the dinner, a woman enters the room, washes Jesus' feet with her tears, dries them with her hair and anoints Him with perfume. Details vary from one gospel to another but the occurrence is so unusual that I believe that they must be four different accounts of the same incident. As with the variation in detail between the four accounts of Jesus' passion and resurrection, the relatively minor differences here lend authenticity since an attempt to construct a false story would result in closer agreement.

Matthew tells us that the house was in Bethany, as do Mark and John, and both Matthew and Mark tell us that the host was a Pharisee known as Simon the Leper. Bethany was a village just a few kilometres east of Jerusalem. Mark and John place the event in the week before Passover. John records that Jesus arrived in Bethany six days before the Passover and that this incident occurred on the day before Jesus' triumphal entry into Jerusalem. Mark tells us that it happened two days before the Passover.

All four gospels use the word μυρος (muros) to describe the substance used by the woman. As mentioned earlier, μυρος (muros) can mean either perfume or ointment. Mark and John specify that it was pure spikenard, and John gives the volume as equivalent of about half a litre. John also remarks that the smell filled the whole house. Anyone who has smelt a sample of spikenard will know that it has a powerful and diffusive odour and such a volume would be more than sufficient to fill a house. Matthew and Mark both say that the ointment was in an alabastron. As explained in Chapter 4, Section 4.2, this could have been made of alabaster or another material since the term is used for any bottle containing either perfume or ointment. The use of the words μυρος (muros) and ναρδος (nardos) together, as also explained above, suggests a concrete of spikenard. Matthew and Mark write that Mary used it to anoint Jesus' head, whereas Luke and John say it was applied to His feet. The Victorian East window in the parish church of Saint Mary, Crundale, Kent, shows Mary pouring perfume out of a bottle onto Jesus' head (Figure 9.2). Anointing His head would be in keeping with the accounts of Matthew and Mark but the free-flowing liquid depicted would be inconsistent with the term μυρος (muros) used in all four gospels and strongly implying a concrete rather than a perfumed oil. Thus, much later interpretation can serve to add to confusion in understanding the original.

Figure 9.2   "Mary Magdalene" window from the church of Saint Mary, Crundale, Kent.

9.3   WHO WAS THE WOMAN?

The identity of the woman is an interesting question. Matthew and Mark just write "a woman", Luke tells us that it was a woman who had led a sinful life in that town. John identifies the woman as Mary. He also says that Martha served the meal and that Lazarus was a guest, reclining at the table with Jesus and the others. It is in his account of the resurrection of Lazarus that John confirms that this Mary is the same as the sister of Martha and Lazarus (John, 11.2).

Mary was a very common name for women in the Holy Land in those days. The name appears 54 times in the New Testament. In 12 of these cases, the name is qualified by the word "of Magdala" but in the other 42 cases, apart from the nativity account in Luke's gospel, there is little detail to indicate which

Mary is intended. Scholars are divided as to how many women called Mary are actually represented by all of these mentions and whether or not any of them are referred to in different ways. For example, Matthew (Matthew, 27.56) and Mark (Mark, 15.40) tell us that Mary mother of James and Joses (Joses is a variant of the name Joseph) were present at Jesus' crucifixion; Mark (Mark, 15.47) says she was present at His entombment; and both Mark (Mark, 16.1) and Luke (Luke, 24.10) record her presence at His resurrection. We know that Jesus had brothers called James and Joses (Matthew, 13.55; Mark, 6.3) so is that Mary the same as the mother of Jesus? Only John (John, 19.25–27) states that Jesus' mother was present at the crucifixion, and so it is possible that John describes her in one way and the other three evangelists describe her differently. Similarly, it is possible that there were two or more women called Mary present at the crucifixion.

The later addition to Mark's gospel says that Jesus had cast seven demons out of Mary of Magdala (Mark, 16.9). Luke (Luke, 8.2) also records this exorcism and that she then joined the group of women who travelled with Jesus from Galilee to Jerusalem. Interestingly, Luke places that note immediately after his account of the supper. Luke's account of the incident suggests that the woman had lived a sinful life in the town of the Pharisee hosting the supper. Magdala was a village on the shore of the Sea of Galilee but the details in Luke's account made it difficult to tell where the supper had taken place and whether the woman's sinful life had been in Magdala or Bethany. Where Luke's account is placed in his gospel would imply that the event took place in the region of Galilee, therefore suggesting Magdala rather than Bethany, and well before Jesus' crucifixion. However, the gospel accounts are not necessarily in strict chronological order and so it could be that Luke wrote that episode out of time with other events. However, despite these uncertainties, the similarities were close enough for Pope Saint Gregory the Great to propose that Mary of Magdala and the Mary, sister of Martha and Lazarus, who lived at Bethany were one and the same person. Similarly, it has been suggested that the woman caught in adultery (John, 7.53–8.11) is also the same person. Adultery in those days often meant prostitution so a tradition developed in Western Christianity that the Mary, sister of Martha and Lazarus, who anointed Jesus at Bethany was Mary of Magdala (Mary

Magdalene), a reformed prostitute and possibly the woman described in John's gospel as the one caught in the guilty act. This teaching is no longer doctrine of the Roman Catholic Church but many Western Christians still hold to it. Eastern Christianity considered Mary of Bethany and Mary of Magdala to be different women, one the sister of Martha and Lazarus and the other a follower from Galilee who had led a saintly life, although this latter is not consistent with the comments about her in the gospels of Mark and Luke. My conclusion is that we cannot know for certain whether or not the woman in Bethany was Mary of Magdala, but John does clearly identify her as the sister of Martha and Lazarus, so I will just call her Mary.

9.4   CONVERSATIONS OVER DINNER

The first three, Matthew, Mark and Luke, of the four gospel records are often referred to as the **synoptic gospels** because they use a similar pragmatic style and record similar types of events. John's gospel is much more contemplative and concentrates on teaching and theology rather than simple recording of events. So one fascinating fact about the accounts of the supper at Bethany is that, in that case, Matthew, Mark and John all give similar accounts whereas Luke records a rather different conversation that took place.

   The gospels of Matthew, Mark and John all tell of a discussion among the disciples about the value of the perfume that Mary used to anoint Jesus. Mark tells us that the value was equivalent to a year's wages. John says it was a pint (about 500 mL) of pure spikenard. Such a quantity today would cost about $500 US dollars. This is less than the annual wages of someone living in North America or Western Europe today, but a survey at the start of the Millennium reported that the average salary in the Republic of Moldavia was about $500 US dollars. This was confirmed to me by a contact in the Moldavian Academy of Sciences in Kishinev. So, for many people nowadays, a pint of pure spikenard still equates to a year's wages. They said that the perfume could have been sold and the money given to the poor. The three gospel accounts all record that Jesus reprimanded the disciples for this conversation and said that the woman had been led to do this in preparation for His burial and that her action would be told everywhere that the

Figure 9.3 "Mary Magdalene" detail from the tympanum of the narthex of Vézelay Abbey.

gospel was preached. Mark and John are clear that it was this event that finally prompted Judas Iscariot to go to the high priests and offer to betray Jesus. John is particularly hard on Judas. He adds that Judas was motivated by greed since he was keeper of the disciples' purse and used to steal from it, implying that had the perfume been sold, it would have added to Judas' personal gains. The scene depicted in the **tympanum** of the **narthex** of Vézelay Abbey in Burgundy (Figure 9.3) shows Mary at Jesus' feet and behind her, on the right, is Judas Iscariot clutching the money bag, reminding us of John's assessment of him. John also notes that many people had come to Bethany to see Lazarus because he was the one Jesus had raised from the dead and that this led the Jewish authorities to plot to kill Lazarus as well as Jesus.

Luke tells us about a very different conversation. His account records that the host was thinking to himself that if Jesus really was a prophet, he would know what sort of woman this was and would not let her touch Him. Jesus, aware of the Pharisee's thoughts, told him a story about two people whose debts to a money lender were both cancelled. One owed a great deal, the other only a small amount. When asked, Simon correctly replied that the one who would show most gratitude was the one who had had the larger debt. Jesus then upbraided Simon for having failed in his duty as a host to provide water for Jesus to wash His feet (the duty of a host whose guests would arrive with feet dusty from wearing sandals when travelling on dusty roads) and did not greet Him with the traditional kiss. Mary had carried out the actions that Simon, the host, should have done. Jesus then pointed out that Mary loved Him and was deeply grateful for what He had done for her. Jesus thus showed Simon that His forgiveness is there for all, whether their debts are small or large. If Mary was a reformed prostitute, then her gift of spikenard was a token of true repentance since she was giving away the tool of her former trade. Repentance does not just mean saying sorry, it means a deliberate decision to change and reform.

So, why is Luke's account different from the other three? It is characteristic of Luke that he tends not to name individuals when doing so could be embarrassing to them. Perhaps it was his training as a medical practitioner that led him to develop that habit. My personal theory is that Luke's account is different because he was the only evangelist who was not present at the dinner. We know from Paul's letter to the church at Colosse (Colossians, 4.10–14) that Luke was a doctor and a Gentile. Simon the Pharisee would not have invited a Gentile to a supper at his house since that would have broken Jewish law. Matthew and John were members of the inner core of 12 disciples. Mark was not in that group, but he does use a literary trick of Greek authors to identify himself as being present at Jesus' arrest in the Garden of Gethsemane. I first came across this device when reading Xenophon's account of the Persian Wars in my Greek class at school. In it, Xenophon describes himself as ο τις Ξενοφον (ho tis Xenophon), a certain man called Xenophon. The Greek text of Mark, 14.51, starts "Και νεανισκος τις" (kai neaniskos tis) meaning "A certain young man", but this construction indicates that young man is the author. So, if Mark was present in Gethsemane, it is quite possible that he was also present at the supper in Bethany. Indeed, it is possible that he is the person that Peter refers to as his son (1 Peter, 5.13). Matthew, Mark and John would thus have all been intent on the discussion about the cost of the spikenard and perhaps not even aware of the conversation that Luke recorded. There is a second reason that we know that Luke would definitely not have been at the meal. We know from the book of Acts that he joined Paul's mission much later, and it would appear that he lived in, or near, Troas before joining Paul. He did visit the Holy Land with Paul at a later date. In the prologue to his gospel (Luke, 1.1–4), Luke tells his reader, Theophilus, that he researched the events of Jesus' life and set down a careful account of them, having spoken with those involved as eye witnesses. Therefore, he must have learned about the supper at Bethany from someone who was there, perhaps Mary or Simon. Since his account presents Mary in a more favourable light than Simon, one is tempted to think that Mary was his source. If so, that might account for Luke seeming to put his version of the supper in a different place and at a different time owing to a misunderstanding of his conversation with Mary. The account (John, 7.53–8.11) of the woman taken in adultery is not present in the oldest copies of John's gospel, such as the Codex Sinaiticus and must have been added at a later date. It is more in the style of Luke than of John, so one might speculate that Luke did write it, that it was lost for some time and then inserted into the wrong gospel. Mary might have told him about that incident at the same time that she told him about the supper at Bethany. That is pure speculation on my part and we cannot know whether it is the case or not.

9.5 DEPICTION IN GLASS

The supper at Bethany is often portrayed in stained glass windows. In the "Mary Magdalene" window in Chartres Cathedral (Figure 9.4), there is a panel showing Mary wiping Jesus' feet

Figure 9.4 A panel from the "Mary Magdalene" window in Chartres Cathedral.

with her hair and the alabastron of spikenard is there at the border. The two figures seated next to Jesus are wearing pointed hats. This is a medieval artistic device to indicate Jewishness of the wearers. So, the artist who made that window had Luke's gospel in mind. The window in the "Mary Magdalene" chapel (Figure 9.5) in the crypt of Canterbury Cathedral shows haloes around the heads of the two people seated next to Jesus (Figure 9.6). The artist in that case must have been thinking of one of the other gospel accounts. In Bourges Cathedral, the corresponding window (Figure 9.7) shows both gospel accounts. Jesus is shown seated at both ends of the table with a Mary at his

Figure 9.5   The "Mary Magdalene" chapel in the crypt of Canterbury Cathedral.
Reproduced with permission from the Dean and Chapter of Canterbury.

Figure 9.6 The window in the "Mary Magdalene" chapel in the crypt of Canterbury Cathedral.
Reproduced with permission from the Dean and Chapter of Canterbury.

Figure 9.7 A panel from the "Mary Magdalene" window in Bourges Cathedral.

feet in both of them. On the left, we see a Jew (Simon?) seated next to Jesus and on the right a disciple.

If the tradition that Mary was a reformed prostitute is correct, then it raises an interesting and somewhat controversial question. Could arguably the most significant event in human history have been triggered by a prostitute opening a bottle of perfume?

Bible References to Perfume, Odour and the Sense of Smell

Genesis, 2.12; Genesis, 8.21; Genesis, 27.27; Genesis, 37.25; Genesis, 43.11; Exodus, 7.21; Exodus, 16.20; Exodus, 25.6; Exodus, 29.18; Exodus, 29.25; Exodus, 29.41; Exodus, 30.1; Exodus, 30.7; Exodus, 30.8; Exodus, 30.9; Exodus, 30.23; Exodus, 30.24; Exodus, 30.25; Exodus, 30.27; Exodus, 30.33; Exodus, 30.34; Exodus, 30.35; Exodus, 30.37; Exodus, 30.38; Exodus, 31.8; Exodus, 31.11; Exodus, 35.8; Exodus, 35.15; Exodus, 35.28; Exodus, 37.25; Exodus, 37.29; Exodus, 39.38; Exodus, 40.5; Exodus, 40.27; Leviticus, 1.9; Leviticus, 1.13; Leviticus, 1.17; Leviticus, 2.1; Leviticus, 2.2; Leviticus, 2.3; Leviticus, 2.12; Leviticus, 2.15; Leviticus, 2.16; Leviticus, 3.5; Leviticus, 3.16; Leviticus, 4.7; Leviticus, 4.31; Leviticus, 5.11; Leviticus, 6.15; Leviticus, 6.21; Leviticus, 8.21; Leviticus, 8.28; Leviticus, 10.1; Leviticus, 16.12; Leviticus, 16.13; Leviticus, 17.6; Leviticus, 23.13; Leviticus, 23.18; Leviticus, 24.7; Leviticus, 26.30; Leviticus, 26.31; Numbers, 4.16; Numbers, 5.15; Numbers, 7.14; Numbers, 7.20; Numbers, 7.26; Numbers, 7.32; Numbers, 7.38; Numbers, 7.44; Numbers, 7.50; Numbers, 7.56; Numbers, 7.62; Numbers, 7.68; Numbers, 7.74; Numbers, 7.80; Numbers, 7.86; Numbers, 16.7; Numbers, 16.17; Numbers, 16.18; Numbers, 16.35; Numbers, 16.40; Numbers, 16.46; Numbers, 16.47; Numbers, 24.6; Deuteronomy, 33.10; Ruth, 3.3; 1 Samuel, 2.28; 1 Samuel, 8.13;

Perfume in the Bible
By Charles Sell
© Charles Sell 2019
Published by the Royal Society of Chemistry, www.rsc.org

1 Kings, 3.3; 1 Kings, 9.25; 1 Kings, 11.8; 1 Kings, 22.43; 2 Kings, 12.3; 2 Kings, 14.4; 2 Kings, 15.4; 2 Kings, 15.35; 2 Kings, 16.4; 2 Kings, 17.11; 2 Kings, 18.4; 2 Kings, 22.17; 2 Kings, 23.5; 2 Kings, 23.8; 1 Chronicles, 6.49; 1 Chronicles, 9.29; 1 Chronicles, 28.18; 2 Chronicles, 2.4; 2 Chronicles, 13.11; 2 Chronicles, 14.5; 2 Chronicles, 16.14; 2 Chronicles, 26.16; 2 Chronicles, 26.18; 2 Chronicles, 28.4; 2 Chronicles, 29.7; 2 Chronicles, 29.11; 2 Chronicles, 30.14; 2 Chronicles, 34.4; 2 Chronicles, 34.7; 2 Chronicles, 34.25; Nehemiah, 3.8; Nehemiah, 13.5; Nehemiah, 13.9; Esther, 2.12; Psalm 45.8; Psalm 115.6; Psalm 141.2; Proverbs, 7.17; Proverbs, 27.9; Ecclesiastes, 7.1; Ecclesiastes, 10.1; Song of Songs, 1.3; Song of Songs, 1.12; Song of Songs, 1.13; Song of Songs, 2.13; Song of Songs, 3.6; Song of Songs, 4.6; Song of Songs, 4.10; Song of Songs, 4.11; Song of Songs, 4.13; Song of Songs, 4.14; Song of Songs, 4.16; Song of Songs, 5.1; Song of Songs, 5.5; Song of Songs, 5.13; Song of Songs, 7.8; Song of Songs, 7.13; Isaiah, 1.13; Isaiah, 3.20; Isaiah, 3.24; Isaiah, 17.8; Isaiah, 27.9; Isaiah, 43.23; Isaiah, 43.24; Isaiah, 57.9; Isaiah, 60.6; Isaiah, 65.3; Isaiah, 66.3; Jeremiah, 1.16; Jeremiah, 6.20; Jeremiah, 7.9; Jeremiah, 11.12; Jeremiah, 11.13; Jeremiah, 11.17; Jeremiah, 17.26; Jeremiah, 18.15; Jeremiah, 19.13; Jeremiah, 32.29; Jeremiah, 41.5; Jeremiah, 44.3; Jeremiah, 44.5; Jeremiah, 44.8; Jeremiah, 44.15; Jeremiah, 44.17; Jeremiah, 44.18; Jeremiah, 44.21; Jeremiah, 44.23; Jeremiah, 44.25; Jeremiah, 48.35; Ezekiel, 6.4; Ezekiel, 6.6; Ezekiel, 6.13; Ezekiel, 8.11; Ezekiel, 16.18; Ezekiel, 16.19; Ezekiel, 20.28; Ezekiel, 20.41; Ezekiel, 23.41; Ezekiel, 27.19; Daniel, 2.46; Hosea, 2.13; Hosea, 11.2; Hosea, 14.6; Habakkuk, 1.16; Malachi, 1.11; Matthew, 2.11; Matthew, 26.7; Matthew, 26.9; Matthew, 26.12; Mark, 14.3; Mark, 14.4; Mark, 14.8; Mark, 15.23; Luke, 1.9; Luke, 1.10; Luke, 1.11; Luke, 7.37; Luke, 7.38; Luke, 7.46; Luke, 23.56; John, 11.2; John, 12.3; John, 12.5; John, 12.7; John, 19.39; 1 Corinthians, 12.17; 2 Corinthians, 2.14; 2 Corinthians, 2.16; Ephesians, 5.2; Philippians, 4.18; Hebrews, 9.4; Revelation, 5.8; Revelation, 8.3; Revelation, 8.4; Revelation, 18.13.

APPENDIX 2

The Early Medieval Windows of Canterbury Cathedral

Many of the illustrations in this book are of windows in Canterbury Cathedral, and so a short note about these windows is appropriate. Canterbury Cathedral has the largest collection of early medieval glass art in the British Isles, despite the damage done by the Commissioners of Henry VIII in the 1530s and 1540s, the Parliamentary army and Puritan iconoclasts of the 1640s and the Nazi bombing raids of the 1940s.

On 5th September 1174, a fire in the city spread to the cathedral and destroyed the roof and interior of the quire built less than a century earlier under the direction of Archbishop Saint Anselm and Priors Ernulf and Conrad. This disaster was turned to advantage since the rebuilding, by Master Masons Guillaume de Sens and his successor William the Englishman, included extension of the cathedral eastwards to include a new Trinity Chapel and Corona Chapel. On 29th December 1170, Archbishop Thomas Becket had been murdered in the cathedral, and two monks, William of Canterbury and Benedict of Peterborough, kept a record of miracles that were reported by locals and by pilgrims who began to arrive in large numbers to visit his tomb that was in a small chapel in the crypt. These records were used to persuade the pope to canonise the martyr and

Perfume in the Bible
By Charles Sell
© Charles Sell 2019
Published by the Royal Society of Chemistry, www.rsc.org

he became Saint Thomas of Canterbury in a relatively short space of time, the bill of canonisation being issued on 21st February 1173. The new Trinity Chapel allowed a shrine to be built in which his remains could be placed and which would help the monks to manage the flow of pilgrims. The records of William and Benedict were also used by the glaziers who made a series of 12 windows illustrating the life of the martyr and his posthumous miracles. These were placed in Trinity Chapel and a series of 12 typological windows were made for the restored quire. These windows could be used by the monks to teach the scriptures and to show how Old Testament events (types) foreshadowed those of the New Testament (anti-types). The "Redemption Window" (made between 1180 and 1207) in the Corona Chapel features prominently in this book and is another example of a typological window. Above the Miracle windows and the Typological windows, running around the clerestory of both the Quire and Trinity Chapel, there was a series, known as "The Ancestors". These windows depicted the ancestors of Jesus according to the genealogy given in Luke's gospel (Luke, 3.23–38) with a few characters added from Matthew's genealogy (Matthew, 1.1–17) to make up the 86 panels needed to fill the available spaces. Jesse trees, showing selected ancestors between Jesse and Jesus (important because of Old Testament prophecies that the Messiah or Christ would be a descendant of Jesse and his son King David) are common in medieval churches and cathedrals but a complete genealogical series is unique to Canterbury. The first of the series depicts Adam Delving (delving = digging) and it is considered that this panel was made in about 1176. That would make it the oldest glass artwork in Britain, although it is possible that some of the ancestor windows might have been salvaged from the fire and reshaped to fit the new window frames, in which case these would be even older.

Some of the glass has been lost over time through the action of the weather, neglect and deliberate destruction by humans. Over half of the glass in the Miracle windows survives, 43 of the original 86 Ancestor windows survive but, of the 12 typological Quire windows, only enough glass has survived to fill two of them. Conservation work started in the 19th century and is still a major activity. George Austin Junior, cathedral glazier

1848–1862, made copies of many of the windows that were lost or beyond repair, including the Jesse trees in the South-East transept and the Corona Chapel. Of the ancestor panels that are currently in the original location of the clerestories of the quire and Trinity Chapel, only eight are originals and 78 are the work of George Austin Junior. Of those other originals of the Ancestors that still survive, 13 (including Adam Delving) are currently in the West window and the remainder are in the South window. Various other original medieval windows, such as the Corona Chapel Redemption Window that features strongly in this book, can be found elsewhere in the cathedral, and there are also some medieval windows, such as the 13th century Mary Magdalene window in the crypt, which were imported from other churches or acquired from collections, notably the Randolph Hearst collection.

Although we tend to describe all glass art as "stained glass", this is not strictly correct since the technique of silver staining was only developed right at the end of the 13th century. Therefore, all of the Canterbury windows used as illustrations here pre-date staining. That technique involved coating glass with silver oxide or a silver salt and heating to allow silver ions to migrate into the glass and stain it yellow. The medieval windows pictured here were made from what is called pot glass.

Glass is made by heating sand (silica) with an alkali. Glass making originated in Syria, where sand was heated with ash from burnt plant material. This was known as soda ash since the plants used came mostly from the Mediterranean coast or the Dead Sea and were rich in sodium. The technology spread to Venice, where the combination of sand from the Venetian lagoon and soda ash from Syria made a fine quality of glass. Because of the need for furnaces in glass making, the Venetian authorities restricted the industry to the island of Murano so that any fires would not damage the rest of the city, and Murano is still the centre of Venetian glass making. In medieval Northern Europe, it was found that wood ash could be used in place of soda ash and the product became known as Waldglas (forest glass). Waldglas is richer in potassium than is soda glass. The type of sand and ash used, together with the conditions employed in the furnace, all combine to affect the properties of the glass, including the way it weathers when exposed to wind, rain and sun. This can

Figure A2.1   The exterior of one of the windows in Chartres Cathedral.

be seen in Figure A2.1, where two different sources of glass were used in one of the windows of Chartres Cathedral, and the differences in the way they have weathered are clearly seen on the exterior.

Early in history, it was found that clear glass could be coloured by adding metal salts or oxides to the molten glass in the furnace. The metal ions become trapped in the glass matrix and are responsible for the colour. Since the ions are held in the glass and away from oxygen or other substances with which they could react, once they have absorbed the energy of ultraviolet or visible light, the only way they can lose that energy is by emitting radiation of a lower frequency, such as infrared light or by warming the glass around them. Thus, the colours do not fade and we can still admire them after 800 years.

Chromium salts were used to give a deep green colour, copper for pale blue and cobalt for deep blue. Depending on the type of iron compound used and the oxidation state of the iron contained in it, iron could be used to produce green, brown or yellow. Metallic gold or, more commonly, metallic copper combined with molten glass gave a red colour so intense that it could not be used directly and a technique known as flashing was developed. In this technique, a thin layer of intensely red glass was fused (flashed) onto clear glass or glass of another colour to add red to it. This also allowed for shading since the red layer could be partially scraped off in places to lighten the red colour.

Using standard glass blowing techniques, a lump of pot glass would be blown into a bottle shape. While the glass was still partially molten and soft, the "bottle" would be cut and flattened out to give a flat piece of glass that could then be cut to the shape required. Fine shaping was done with a tool known as a grozing iron that was used to chip off tiny fragments from the edge of the piece.

Apart from flashed glass, each piece of glass used in the 12th and early 13th centuries was of a uniform single colour. Following the outline of the artist's drawing, a piece of glass would be cut to the shape needed for each section of the painting. In order to add detail, such as facial features on a flesh coloured piece of glass, a layer of black glass was fused onto the coloured sections and then scraped off where not wanted. The pieces were then assembled into panels using lead strips to hold the fragments together. This accounts for the leading patterns that are an intrinsic part of medieval glass art. (Later repairs sometimes result in additional strips of lead being used where the original was a single piece of glass.) The panels of glass painting were then mounted on an iron framework called an armature and held in place by iron pegs. These details can also be seen in Figure A2.1. The armatures were treated with oil rather than paint, and time has shown the wisdom of this approach since water tends to become trapped between a layer of paint and the metal surface, resulting in rusting. Similarly, in the best examples of window construction, the armatures were made to sit into the window frame and be held in place by mortar, as seen in Figure A2.1. In cases where iron pins on the armature projected into the stonework, it has been found that, over the centuries, rusting of the pins causes the stone to crack and fall apart.

Nowadays, when medieval glass is restored, an outer layer of clear modern glass is often added to protect the precious medieval glass from further weathering. A technique used at Canterbury Cathedral since the 1970s places the protective glass in leading patterns matching those of the medieval glass so that when viewed from outside, it is very difficult to tell that one is

Figure A2.2   Protective modern glass for a Miracle window.

Figure A2.3   External view of a protected medieval window of Canterbury Cathedral.
Reproduced with permission from the Dean and Chapter of Canterbury.

not looking directly at the medieval glass. Figure A2.2 shows the protective glass in place waiting for one of the Miracle windows to be replaced after restoration. The glaziers' scaffolding can be seen in the foreground. Figure A2.3 shows the external view of another of Canterbury's medieval windows, illustrating the success of the technique.

Bibliography

Recommended sources for further reading are listed below in alphabetical order of subject groups and in chronological order of publication within each group.

BIBLICAL STUDIES

There are two websites that I have found of particular use when searching for Biblical material or checking Hebrew, Greek and English equivalents or translations.

Biblehub contains the original Hebrew of the Old Testament and Greek of the New Testament with translations into many different modern languages. The interlinear option shows the original language with the English alongside. Users just have to remember that, whilst English reads left to right, Hebrew reads from right to left. The URL for the site is as follows:

https://biblehub.com/interlinear

The Septuagint can be found *via* the ellopos website where the original Greek and an English translation are found side by side. The URL for the site is as follows together with a link that takes one directly to the Septuagint:

https://www.ellopos.net
https://www.ellopos.net/elpenor/greek-texts/septuagint/default.asp

Perfume in the Bible
By Charles Sell
© Charles Sell 2019
Published by the Royal Society of Chemistry, www.rsc.org

Ancient manuscripts were copied by hand and so copyists' errors, editorial modifications, insertions and deletions all mean that various copies differ from one another in places. Readers using these websites should remember that the single line of text is one judged by its editors to best represent the original message. The oldest known complete copy of the Christian Bible is the Codex Sinaiticus, which was written about 330–360 AD and was discovered in the monastery of Saint Catherine on the Sinai Peninsula. The original Codex can be viewed on the Codex Sinaiticus website and even a cursory inspection will reveal the changes made to it over time. The URL for the site is as shown below:

http://codexsinaiticus.org/en/

The SPCK Bible Atlas, ed. B. McClenahan, SPCK, London, 2013, ISBN 978-0-281-06851-7

The Oxford Companion to the Bible, ed. B. M. Metzger and M. D. Coogan, Oxford University Press, Oxford, 1993, ISBN 10: 0-19-504645-5

CHEMISTRY RELATED TO PERFUME

Chapters 6, 7 & 8 in C. S. Sell, *Chemistry and the Sense of Smell*, John Wiley & Sons Inc., Hoboken, New Jersey, 2014, ISBN 978-0-470-55130-1

C. S. Sell, Terpenoids, in *Kirk-Othmer Encyclopedia of Chemical Technology*, vol. 22, Wiley, Hoboken, New Jersey, 2007.

C. S. Sell, Ingredients for the Modern Perfumery Industry, in *The Chemistry of Fragrance*, ed. C. S. Sell, Royal Society of Chemistry, Cambridge, 2nd edn, 2006, ISBN: 978-0-85404-824-3

K. Jenner, The Search for New Ingredients, in *The Chemistry of Fragrance*, ed. C. S. Sell, Royal Society of Chemistry, Cambridge, 2nd edn, 2006, ISBN: 978-0-85404-824-3

H. Surburg and J. Panten, *Common Fragrance and Flavour Materials*, 5th edn, Wiley VCH, Weinheim, 2006.

C. S. Sell, *A Fragrant Introduction to Terpenoid Chemistry*, Royal Society of Chemistry, Cambridge, 2003.

G. Ohloff, *Scent and Fragrances*, Springer-Verlag, Berlin, 1994.

Perfumes, Art, Science and Technology, ed., P. M. Müller and D. Lamparsky, Elsevier, London, 1991.

S. Arctander, *Perfume and Flavour Chemicals*, 2 vol., S. Arctander, Elizabeth, New Jersey, 1969.

Journal Review Articles

C. S. Sell, On the unpredictability of odour, *Angew. Chem. Int. Edn.*, 2006, **45**(38), 6254.

P. Kraft, J. Bajgrowicz, C. Denis and G. Fráter, Odds and trends: recent developments in the chemistry of odorants, *Angew. Chem. Int. Edn.*, 2001, **39**(17), 2981.

G. Fráter, J. Bajgrowicz and P. Kraft, Fragrance chemistry, *Tetrahedron*, 1998, **54**(27), 7633.

K. J. Rossiter, Structure-odour relationships, *Chem. Rev.*, 1996, **96**(8), 3201.

ESSENTIAL OILS AND OTHER NATURAL PERFUME INGREDIENTS

The two classical books on essential oils are both now out of print, but if copies can be found through a library, they are well worth studying. Both are very authoritative multi-volume books.

E. Günther, *The Essential Oils*, D. Van Nostrand & Co., New York, 1948.

K. Gildemeister and Fr. Hoffmann, *Die Ætherischen Öle*, Akademie Verlag, Berlin, 1959.

Handbook of Essential Oils, ed. K. H. Can Baser and G. Buchbauer, CRC Press (Taylor and Francis), Boca Raton, Florida, 2nd edn, 2016, ISBN: 9781466590465

Chapter 5 in C. S. Sell, *Chemistry and the Sense of Smell*, John Wiley & Sons Inc., Hoboken, New Jersey, 2014, ISBN: 978-0-470-55130-1

L. J, Musselman, *Figs, Dates, Laurel and Myrrh, Plants of the Bible and the Quran*, Timber Press, Portland, Oregon, 2007, ISBN: 978-0-88192-855-6

C. S. Sell, Perfumery Materials of Natural Origin, in *The Chemistry of Fragrance*, ed. C. S. Sell, Royal Society of Chemistry, Cambridge, 2nd edn, 2006, ISBN: 978-0-85404-824-3

R. Clery, Natural Product Analysis in the Fragrance Industry, in *The Chemistry of Fragrance*, ed. C. S. Sell, Royal Society of Chemistry, Cambridge, 2nd edn, 2006, ISBN: 978-0-85404-824-3

R. Kaiser, *Meaningful Scents around the World*, Wiley-VCH, Weinheim, 2006.

C. S. Sell, *A Fragrant Introduction to Terpenoid Chemistry*, Royal Society of Chemistry, Cambridge, 2003.

D. G. Williams, *The Chemistry of Essential Oils*, Micelle Press, Weymouth, 1996, ISBN: 978-1-870228-12-1.

J. Mann, R. S. Davidson, J. B. Hobbs, D. V. Banthorpe and J. B. Harbourne, *Natural Products: Their Chemistry and Biological Significance*, Longman, Harlow, 1994.

S. Arctander, *Perfume and Flavour Materials of Natural Origin*, S. Arctander, Elizabeth, New Jersey, 1960.

OLFACTION, HOW THE SENSE OF SMELL WORKS

Chapters 1 & 2 in C. S. Sell, *Chemistry and the Sense of Smell*, John Wiley & Sons Inc., Hoboken, New Jersey, 2014, ISBN: 978-0-470-55130-1.

G. M. Shepherd, *Neurogastronomy*, Columbia University Press, New York, 2012, ISBN: 978-0-231-15910-4.

R. L. Doty, *The Great Pheromone Myth*, The Johns Hopkins University Press, Baltimore, 2010, ISBN: 978-0-8018-9347-6.

C. S. Sell, Olfaction, in *The Wiley Encyclopedia of Chemical Biology*, Wiley, Hoboken, 2008.

D. A. Wilson and R. J. Stevenson, *Learning to Smell*, The Johns Hopkins University Press, Baltimore, 2006.

C. S. Sell, Chemoreception, in *The Chemistry of Fragrance*, ed. C. S. Sell, Royal Society of Chemistry, Cambridge, 2nd edn, 2006, ISBN: 978-0-85404-824-3.

P. Jellinek, *The Psychological Basis of Perfumery*, Blackie Academic and Professional, London, 1997, ISBN 978-0-7514-0368-8.

S. van Toller and G. H. Dodd, ed., *Perfumery, The Psychology and Biology of Fragrance*, Chapman and Hall, London, 1988, ISBN 0-412-30010-9.

W. James, *The Principles of Psychology*, Harvard University Press, Cambridge MA, 1983.

W. James, *The Principles of Psychology*, Dover Publications, 1950.

W. James, *The Principles of Psychology*, Henry Holt and Company, New York, 1890.

PERFUMERY

L. Harman, The Human Relationship with Fragrance, in *The Chemistry of Fragrance*, ed. C. S. Sell, Royal Society of Chemistry, Cambridge, 2nd edn, 2006, ISBN: 978-0-85404-824-3

D. H. Pybus, The History of Aroma Chemistry and Perfume, in *The Chemistry of Fragrance*, ed. C. S. Sell, Royal Society of Chemistry, Cambridge, 2nd edn, 2006, ISBN: 978-0-85404-824-3

D. H. Pybus, The Structure of an International Fragrance Company, in *The Chemistry of Fragrance*, ed. C. S. Sell, Royal Society of Chemistry, Cambridge, 2nd edn, 2006, ISBN: 978-0-85404-824-3

D. H. Pybus, The Perfume Brief, in *The Chemistry of Fragrance*, ed. C. S. Sell, Royal Society of Chemistry, Cambridge, 2nd edn, 2006, ISBN: 978-0-85404-824-3

L. Small, Perfumer Creation: The Role of the Perfumer, in *The Chemistry of Fragrance*, ed. C. S. Sell, Royal Society of Chemistry, Cambridge, 2nd edn, 2006, ISBN: 978-0-85404-824-3

Churchill, Measurement of Fragrance Perception, in *The Chemistry of Fragrance*, ed. C. S. Sell, Royal Society of Chemistry, Cambridge, 2nd edn, 2006, ISBN 1: 978-0-85404-824-3

J. Beerling, The Application of Fragrance, in *The Chemistry of Fragrance*, ed. C. S. Sell, Royal Society of Chemistry, Cambridge, 2nd edn, 2006, ISBN: 98-0-85404-824-3

S. Meakins, The Safety and Toxicology of Fragrances, in *The Chemistry of Fragrance*, ed. C. S. Sell, Royal Society of Chemistry, Cambridge, 2nd edn, 2006, ISBN: 98-0-85404-824-3

K. D. Perring, Volatility and Substantivity, in *The Chemistry of Fragrance*, ed. C. S. Sell, Royal Society of Chemistry, Cambridge, 2nd edn, 2006, ISBN: 978-0-85404-824-3

D. H. Pybus, Buying Fragrance Ingredients and Selling Fragrance Compounds, in *The Chemistry of Fragrance*, ed. C. S. Sell, Royal Society of Chemistry, Cambridge, 2nd edn, 2006, ISBN: 978-0-85404-824-3

L. Small, The Finale: Brief Submission, in *The Chemistry of Fragrance*, ed. C. S. Sell, Royal Society of Chemistry, Cambridge, 2nd edn, 2006, ISBN: 978-0-85404-824-3

D. G. Williams, *Perfumes of Yesterday*, Micelle Press, Weymouth, 2004.

M. Edwards, *Fragrances of the World*, Allured Publishing Corp., Carol Stream, IL, 2004.

M. Edwards, *Perfume Legends*, HM Editions, 1996.

T. Curtis and D. G. Williams, *Introduction to Perfumery*, 2nd edn, Ellis Horwood, Hemel Hempstead, 2001, ISBN 9781870228244

R. R. Calkin and J. S. Jellinek, *Perfumery, Practice and Principles*, John Wiley and Sons, New York, 1994, ISBN 0-471-58934-9

Herodotus, Tr. A. de Sélincourt, *The Histories*, Penguin Classics, Harmondsworth, 1954, pp. 220–221.

Biblical References Index

Genesis
2.12   3, 61, 79
9.13   81
19   70
27.27   17–18
37.25   40, 71
43.11   71

Exodus
7.21   29
12.8   40
16.20   29
30.1   38
30.1–3   109
30.1–10   63
30.7–8   109
30.9   109
30.22–38   40, 104–107
30.23   56–57, 73
30.24   55, 80
30.30   117
30.34   65, 70, 76–77, 91, 109
30.34–35   38
30.35   3, 100
30.38   117
40.5   109

Leviticus
2.1   109
2.2   109
10.1–2   107
11.9, 12, 19   78
14.4   39
14.52   40
16.12   63, 109
16.12–13   107, 109–110
20.5   104
20.17   104
26.30–31   111

Numbers
5.15   37
16.35   107
16.40   107
16.47   110
24.6   43, 43–44

Deuteronomy
23.3   108, 124

Judges
5.4   92

Ruth
3.3   123

Subject Index

Page numbers in *italics* indicate the subject is only in an illustration on that page. Other page references may include illustrations. Page numbers in **bold** indicate a glossary definition.